STUDENT SOLUTIONS MANUAL

to accompany

APPLIED CALCULUS

Deborah Hughes-Hallett
Harvard University

Andrew M. Gleason
Harvard University

Patti Frazer Lock
St. Lawrence University

Daniel E. Flath
University of South Alabama

et al.

John Wiley & Sons, Inc.

New York • Chichester • Weinheim • Brisbane • Singapore • Toronto

This project was supported, in part,
by the
National Science Foundation
Opinions expressed are those of the authors
and not necessarily those of the Foundation

COVER PHOTO: Tom Bean/Tony Stone Images

ISBN 0-471-17351-7

Printed in the United States of America

10 9 8 7 6 5 4 3 2 1

Printed and bound by Victor Graphics, Inc.

CONTENTS

CHAPTER ONE

Solutions for Section 1.1

1. Between 1950 and 1995, we have

$$\text{Average rate of change} = \frac{\text{Change in marine catch}}{\text{Change in years}}$$
$$= \frac{91 - 17}{1995 - 1950}$$
$$= \frac{74}{45}$$
$$= 1.64 \text{ million tons/year}$$

Between 1950 and 1995, marine catch increased at an average rate of 1.64 million tons each year.

5. (a) The total change in the public debt during this 13-year period was $4351.2 - 907.7 = 3443.5$ billion dollars.
    (b) The average rate of change is given by

$$\begin{array}{l}\text{Average rate of change} \\ \text{of the public debt} \\ \text{between 1980 and 1993}\end{array} = \frac{\text{Change in public debt}}{\text{Change in time}}$$
$$= \frac{4351.2 - 907.7}{1993 - 1980}$$
$$= \frac{3443.5 \text{ billion dollars}}{13 \text{ years}}$$
$$= 264.88 \text{ billion dollars per year.}$$

The units for the average rate of change are units of public debt over units of time, or billions of dollars per year. Between 1980 and 1993, the public debt of the United States increased at an average rate of 264.88 billion dollars per year. (This represents an increase of \$725,700,000, over 700 million dollars, every day!)

9. (a) Negative. Rain forests are being continually destroyed to make way for housing, industry and other uses.
    (b) Positive. Virtually every country has a population which is increasing, so the world's population must also be increasing overall.
    (c) Negative. Since a vaccine for polio was found, the number of cases has dropped every year to almost zero today.
    (d) Negative. As time passes and more sand is eroded, the height of the sand dune decreases.
    (e) Positive. As time passes the price of just about everything tends to increase.

13. (a) The average rate of change R of the sperm count is

$$R = \frac{66 - 113}{1990 - 1940} = -0.94 \text{ million sperm per milliliter per year.}$$

    (b) We want to find how long it will take 66 million to drop to 20 million, given that annual rate of change is -0.94. We write

$$66 + n(-0.94) = 20$$
$$n(-0.94) = -46.$$

Solving for n gives
$$n \approx 49 \text{ years.}$$

The average sperm count would go below 20 million in 2039.

Solutions for Section 1.2

1. (a) The story in (a) matches Graph (IV), in which the person forgot her books and had to return home.
    (b) The story in (b) matches Graph (II), the flat tire story. Note the long period of time during which the distance from home did not change (the horizontal part).
    (c) The story in (c) matches Graph (III), in which the person started calmly but sped up later.

    The first graph (I) does not match any of the given stories. In this picture, the person keeps going away from home, but his speed decreases as time passes. So a story for this might be: *I started walking to school at a good pace, but since I stayed up all night studying calculus, I got more and more tired the farther I walked.*

5.

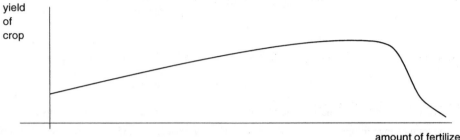

9.

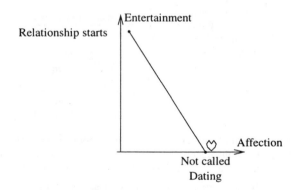

13. (a) Substituting $x = 1$ gives $f(1) = 3(1) - 5 = 3 - 5 = -2$.
    (b) We substitute $x = 5$:
    $$y = 3(5) - 5 = 15 - 5 = 10.$$
    (c) We substitute $y = 4$ and solve for x:
    $$4 = 3x - 5$$
    $$9 = 3x$$
    $$x = 3$$
    (d) Average rate of change $= \dfrac{f(4) - f(2)}{4 - 2} = \dfrac{7 - 1}{2} = \dfrac{6}{2} = 3$.

17. The average rate of change R between $x = 1$ and $x = 3$ is
    $$R = \frac{f(3) - f(1)}{3 - 1}$$
    $$= \frac{18 - 2}{2}$$
    $$= \frac{16}{2}$$
    $$= 8.$$

Solutions for Section 1.3

1. (a) is (V), because slope is positive, vertical intercept is negative
   (b) is (IV), because slope is negative, vertical intercept is positive
   (c) is (I), because slope is 0, vertical intercept is positive
   (d) is (VI), because slope and vertical intercept are both negative
   (e) is (II), because slope and vertical intercept are both positive
   (f) is (III), because slope is positive, vertical intercept is 0

5. $y - c = m(x - a)$

9. For the function given by table (a), we know that the slope is

$$\text{slope} = \frac{27 - 25}{0 - 1} = -2.$$

We also know that at $x = 0$ we have $y = 27$. Thus we know that the vertical intercept is 27. The formula for the function is

$$y = -2x + 27.$$

For the function in table (b), we know that the slope is

$$\text{slope} = \frac{72 - 62}{20 - 15} = \frac{10}{5} = 2.$$

Thus, we know that the function will take on the form

$$s = 2t + b.$$

Substituting in the coordinates $(15, 62)$ we get

$$
\begin{aligned}
s &= 2t + b \\
62 &= 2(15) + b \\
&= 30 + b \\
32 &= b
\end{aligned}
$$

Thus, a formula for the function would be

$$s = 2t + 32.$$

13. (a) We know that the function for q in terms of p will take on the form

$$q = mp + b.$$

We know that the slope will represent the change in q over the corresponding change in p. Thus

$$m = \text{slope} = \frac{4 - 3}{12 - 15} = \frac{1}{-3} = -\frac{1}{3}.$$

Thus, the function will take on the form

$$q = -\frac{1}{3}p + b.$$

Substituting the values $q = 3, p = 15$, we get

$$
\begin{aligned}
3 &= -\frac{1}{3}(15) + b \\
3 &= -5 + b \\
b &= 8.
\end{aligned}
$$

Thus, the formula for q in terms of p is

$$q = -\frac{1}{3}p + 8.$$

(b) We know that the function for p in terms of q will take on the form

$$p = mq + b.$$

We know that the slope will represent the change in p over the corresponding change in q. Thus

$$m = \text{slope} = \frac{12 - 15}{4 - 3} = -3.$$

Thus, the function will take on the form

$$p = -3q + b.$$

Substituting the values $q = 3, p = 15$ again, we get

$$15 = (-3)(3) + b$$
$$15 = -9 + b$$
$$b = 24.$$

Thus, a formula for p in terms of q is

$$p = -3q + 24.$$

17. By joining consecutive points we get a line whose slope is the average rate of change. The steeper this line, the greater the average rate of change. See Figure 1.1.

(a) C and D. Steepest slope.
     B and C. Slope closest to 0.
(b) A and B, and C and D gives the 2 slopes closest to each other.

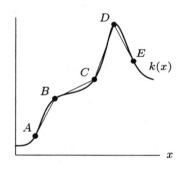

Figure 1.1

21. The average rate of change R from $x = -2$ to $x = 1$ is:

$$R = \frac{f(1) - f(-2)}{1 - (-2)} = \frac{3(1)^2 + 4 - (3(-2)^2 + 4)}{1 + 2} = \frac{7 - 16}{3} = -3$$

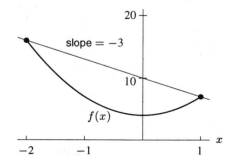

Solutions for Section 1.4

1. (a) We know that regardless of the number of rides one takes, one must pay $7 to get in. After that, for each ride you must pay another $1.50, thus the function $R(n)$ is

$$R(n) = 7 + 1.5n.$$

(b) Substituting in the values $n = 2$ and $n = 8$ into our formula for $R(n)$ we get

$$R(2) = 7 + 1.5(2) = 7 + 3 = \$10.$$

This means that admission and 2 rides costs $10.

$$R(8) = 7 + 1.5(8) = 7 + 12 = \$19.$$

This means that admission and 8 rides costs $19.

5. (a) The cost of producing 500 units is

$$C(500) = 6000 + 10(500) = 6000 + 5000 = \$11{,}000.$$

The revenue the company makes by selling 500 units is

$$R(500) = 12(500) = \$6000.$$

Thus, the cost of making 500 units is greater than the money the company will make by selling the 500 units, so the company does not make a profit.

The cost of producing 5000 units is

$$C(5000) = 6000 + 10(5000) = 6000 + 50000 = \$56{,}000.$$

The revenue the company makes by selling 5000 units is

$$R(5000) = 12(5000) = \$60{,}000.$$

Thus, the cost of making 5000 units is less than the money the company will make by selling the 5000 units, so the company does make a profit.

(b) The break-even point is the number of units that the company has to produce so that in selling those units, it makes as much as it spent on producing them. That is, we are looking for q such that

$$C(q) = R(q).$$

Solving we get

$$C(q) = R(q)$$
$$6000 + 10q = 12q$$
$$2q = 6000$$
$$q = 3000.$$

Thus, if the company produces and sells 3000 units, it will break even.

Graphically, the break-even point, which occurs at $(3000, \$36{,}000)$, is the point at which the graphs of the cost and the revenue functions intersect. (See Figure 1.2.)

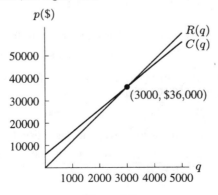

Figure 1.2

9.   (a)   The cost function is of the form

$$C(q) = b + m \cdot q$$

where m is the variable cost and b is the fixed cost. Since the variable cost is \$20 and the fixed cost is \$650,000, we get

$$C(q) = 650{,}000 + 20q.$$

The revenue function is of the form

$$R(q) = pq$$

where p is the price that the company is charging the buyer for one pair. In our case the company charges \$70 a pair so we get

$$R(q) = 70q.$$

The profit function is the difference between revenue and cost, so

$$\pi(q) = R(q) - C(q) = 70q - (650{,}000 + 20q) = 70q - 650{,}000 - 20q = 50q - 650{,}000.$$

(b)   Marginal cost is \$20 per pair. Marginal revenue is \$70 per pair. Marginal profit is \$50 per pair.

(c)   We are asked for the number of pairs of shoes that need to be produced and sold so that the profit is larger than zero. That is, we are trying to find q such that

$$\pi(q) > 0.$$

Solving we get

$$\pi(q) > 0$$
$$50q - 650{,}000 > 0$$
$$50q > 650{,}000$$
$$q > 13{,}000.$$

Thus, if the company produces and sells more than 13,000 pairs of shoes, it will make a profit.

13.   See Figure 1.3

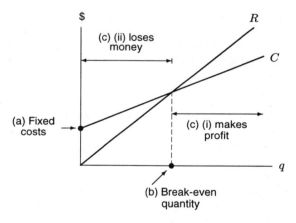

Figure 1.3

17.   (a)   We know that the function for the value of the tractor will be of the form

$$V(t) = m \cdot t + b$$

where m is the slope and b is the vertical intercept. We know that the vertical intercept is simply the value of the function at time $t = 0$, which is \$50,000. Thus

$$b = \$50{,}000.$$

Since we know the value of the tractor at time $t = 20$ we know that the slope is

$$m = \frac{V(20) - V(0)}{20 - 0} = \frac{10{,}000 - 50{,}000}{20} = \frac{-40{,}000}{20} = -2000.$$

Thus we get

$$V(t) = -2000t + 50{,}000 \text{ dollars.}$$

(b)

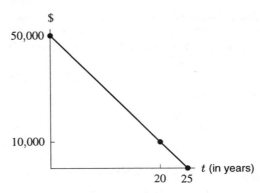

Figure 1.4

(c) Looking at Figure 1.4 we see that the vertical intercept occurs at the point $(0, 50,000)$ and the horizontal intercept occurs at $(25, 0)$. The vertical intercept tells us the value of the tractor at time $t = 0$, namely, when it was brand new. The horizontal intercept tells us at what time t the value of the tractor will be $0. Thus the tractor is worth $50,000 when it is new, and it is worth nothing after 25 years.

21. (a) Looking at the graph we see that it goes through the point $(20, 17)$, that is, there are 20 items bought when the price is $17. Looking at the graph we see that it goes through the point $(50, 8)$, that is, there are 50 items bought when the price is $8.

(b) Looking at the graph we see that it goes through the point $(30, 13)$, that is, at the price of $13 per item, 30 items are bought. Looking at the graph we see that it goes through the point $(10, 25)$, that is, at the price of $25 per item, 10 items are bought.

25. (a) We know that the equilibrium point is the point where the supply and demand curves intersect. Looking at the figure in the problem, we see that the price at which they intersect is $10 per unit and the corresponding quantity is 3000 units.

(b) We know that the supply curve climbs upwards while the demand curve slopes downwards. Thus we see from the figure that at the price of $12 per unit the suppliers will be willing to produce 3500 units while the consumers will be ready to buy 2500 units. Thus we see that when the price is above the equilibrium point, more items would be produced than the consumers will be willing to buy. Thus the producers end up wasting money by producing that which will not be bought, so the producers are better off lowering the price.

(c) Looking at the point on the rising curve where the price is $8 per unit, we see that the suppliers will be willing to produce 2500 units, whereas looking at the point on the downward sloping curve where the price is $8 per unit, we see that the consumers will be willing to buy 3500 units. Thus we see that when the price is less than the equilibrium price, the consumers are willing to buy more products than the suppliers would make and the suppliers can thus make more money by producing more units and raising the price.

29. (a) $k = p_1 s + p_2 l$ where $s = \#$ of liters of soda and $l = \#$ of liters of oil.

(b) If $s = 0$, then $l = \frac{k}{p_2}$. Similarly, if $l = 0$, then $s = \frac{k}{p_1}$. These two points give you enough information to draw a line containing the points which satisfy the equation.

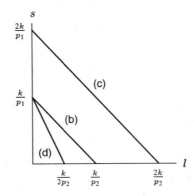

(c) If the budget is doubled, we have the constraint: $2k = p_1 s + p_2 l$. We find the intercepts as before. If $s = 0$, then $l = \frac{2k}{p_2}$; if $l = 0$, then $s = \frac{2k}{p_1}$. The intercepts are both twice what they were before.

(d) If the price of oil doubles, our constraint is $k = p_1 s + 2p_2 l$. Then, calculating the intercepts gives that the s intercept remains the same, but the l intercept gets cut in half. In other words, $s = 0$ means $l = \frac{k}{2p_2} = \frac{1}{2}\frac{k}{p_2}$. Therefore the maximum amount of oil you can buy is half of what it was previously.

33. We know that at the point where the price is \$1 per scoop the quantity must be 240. Thus we can fill in the graph as follows:

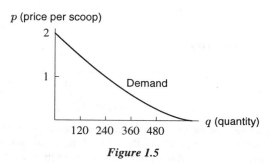

Figure 1.5

(a) Looking at Figure 1.5 we see that when the price per scoop is half a dollar, the quantity given by the demand curve is roughly 360 scoops.

(b) Looking at Figure 1.5 we see that when the price per scoop is \$1.50, the quantity given by the demand curve is roughly 120 scoops.

Solutions for Section 1.5

1. One possible answer is shown in Figure 1.6.

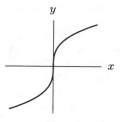

Figure 1.6

5. If distance is d, then $v = \dfrac{d}{t}$.

9. $y = 5x^{1/2}; k = 5, p = 1/2$.

13. $y = 9x^{10}; k = 9, p = 10$.

17. $y = 8x^{-1}; k = 8, p = -1$

21. $y = x^4$ goes to positive infinity in both cases.

25. For $y = x$, average rate of change $= \frac{10-0}{10-0} = 1$.
    For $y = x^2$, average rate of change $= \frac{100-0}{10-0} = 10$.
    For $y = x^3$, average rate of change $= \frac{1000-0}{10-0} = 100$.
    For $y = x^4$, average rate of change $= \frac{10000-0}{10-0} = 1000$.
    So $y = x^4$ has the largest average rate of change. For $y = x$, the line is the same as the original function.

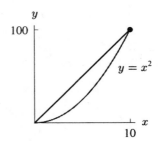

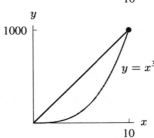

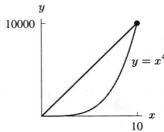

29. Let M = blood mass and B = body mass. Then $M = k \cdot B$. Using the fact that $M = 150$ when $B = 3000$, we have

$$M = k \cdot B$$
$$150 = k \cdot 3000$$
$$k = 150/3000 = 0.05$$

We have $M = 0.05B$. For a human with $B = 70$, we have $M = 0.05(70) = 3.5$ kilograms of blood.

33. (a) Note that

$$\frac{17.242}{50} = 0.34484,$$
$$\frac{25.863}{75} = 0.34484,$$
$$\frac{34.484}{100} = 0.34484,$$
$$\frac{51.726}{150} = 0.34484,$$

and finally

$$\frac{68.968}{200} = 0.34484$$

Thus, the proportion of A to m remains constant for all A and corresponding m, and so A is proportional to m.

(b) Since

$$\frac{31.447}{8} \approx 3.931$$

and

$$\frac{44.084}{12} \approx 3.674$$

we see that the proportion of r to m does not remain constant for different values of r and corresponding values of m. Thus, we see that r and m are not proportional.

37. The graph described is shown in Figure 1.7. It most closely resembles the function x^3: it cannot be any of the other four functions mentioned.

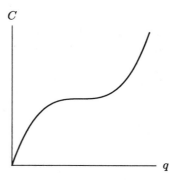

Figure 1.7

41.   The values in Table 1.1 suggest that this limit is 0. The graph of $y = 1/x$ in Figure 1.8 suggests that $y \to 0$ as $x \to \infty$ and so supports the conclusion.

TABLE 1.1

x	100	1000	1,000,000
$1/x$	0.01	0.001	0.000001

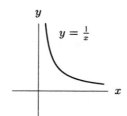

Figure 1.8

Solutions for Section 1.6

1.   (a)   Initial amount $= 100$; exponential growth; growth rate $= 7\% = 0.07$.
     (b)   Initial amount $= 5.3$; exponential growth; growth rate $= 5.4\% = 0.054$.
     (c)   Initial amount $= 3,500$; exponential decay; decay rate $= -7\% = -0.07$.
     (d)   Initial amount $= 12$; exponential decay; decay rate $= -12\% = -0.12$.

5.

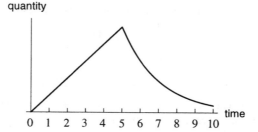

9.   Since 2^x is always positive, the graph of $y = 2^x$ is above the x-axis. The fact that $2^{-x} = 1/2^x$ tells us that 2^{-x} is always positive as well, and that 2^{-x} is small where 2^x is large and vice versa. The graph of $y = 2^{-x}$ is large for negative x and small for positive x. The graphs of $y = 2^x$ and $y = 2^{-x}$ are reflections of one another in the y-axis. (This happens because 2^{-x} is obtained from 2^x by replacing x by $-x$.)

13. (a) In this case we know that
$$f(1) - f(0) = 12.7 - 10.5 = 2.2$$

while
$$f(2) - f(1) = 18.9 - 12.7 = 6.2.$$

Thus, the function described by these data is not a linear one. Next we check if this function is exponential.
$$\frac{f(1)}{f(0)} = \frac{12.7}{10.5} \approx 1.21$$

while
$$\frac{f(2)}{f(1)} = \frac{18.9}{12.7} \approx 1.49$$

thus $f(x)$ is not an exponential function either.

(b) In this case we know that
$$s(0) - s(-1) = 30.12 - 50.2 = -20.08$$

while
$$s(1) - s(0) = 18.072 - 30.12 = -12.048.$$

Thus, the function described by these data is not a linear one. Next we check if this function is exponential.
$$\frac{s(0)}{s(-1)} = \frac{30.12}{50.2} = 0.6,$$
$$\frac{s(1)}{s(0)} = \frac{18.072}{30.12} = 0.6,$$

and
$$\frac{s(2)}{s(1)} = \frac{10.8432}{18.072} = 0.6.$$

Thus, $s(t)$ is an exponential function. We know that $s(t)$ will be of the form
$$s(t) = P_0 a^t$$

where P_0 is the initial value and $a = 0.6$ is the base. We know that
$$P_0 = s(0) = 30.12.$$

Thus,
$$s(t) = 30.12a^t.$$

Since $a = 0.6$, we have
$$s(t) = 30.12(0.6)^t.$$

(c) In this case we know that
$$\frac{g(2) - g(0)}{2 - 0} = \frac{24 - 27}{2} = \frac{-3}{2} = -1.5,$$
$$\frac{g(4) - g(2)}{4 - 2} = \frac{21 - 24}{2} = \frac{-3}{2} = -1.5,$$

and
$$\frac{g(6) - g(4)}{6 - 4} = \frac{18 - 21}{2} = \frac{-3}{2} = -1.5.$$

Thus, $g(u)$ is a linear function. We know that
$$g(u) = m \cdot u + b$$

where m is the slope and b is the vertical intercept, or the value of the function at zero. So
$$b = g(0) = 27$$

and from the above calculations we know that
$$m = -1.5.$$

Thus.
$$g(u) = -1.5u + 27.$$

17.   (a)   Since $P(t)$ is an exponential function, it will be of the form $P(t) = P_0 a^t$. We have $P_0 = 1$, since 100% is present at time $t = 0$, and $a = 0.975$, because each year 97.5% of the contaminant remains. Thus,

$$P(t) = (0.975)^t.$$

(b)   The graph is shown in Figure 1.9.

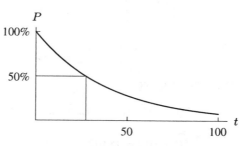

Figure 1.9

(c)   The half-life is about 27 years, since $(0.975)^{27} \approx 0.5$.

(d)   At time $t = 100$ there appears to be about 8% remaining, since $(0.975)^{100} \approx 0.08$.

21.   (a)   The slope is given by

$$m = \frac{P - P_1}{t - t_1} = \frac{100 - 50}{20 - 0} = \frac{50}{20} = 2.5.$$

We know $P = 50$ when $t = 0$, so

$$P = 2.5t + 50.$$

(b)   Given $P = P_0 a^t$ and $P = 50$ when $t = 0$,

$$50 = P_0 a^0, \text{ so } P_0 = 50.$$

Then, using $P = 100$ when $t = 20$

$$100 = 50a^{20}$$
$$2 = a^{20}$$
$$a = 2^{1/20} \approx 1.035265.$$

And so we have

$$P = 50(1.035265)^t.$$

The completed table is found in Table 1.2.

(c)

TABLE 1.2   *The cost of a home*

	(a) Linear Growth	(b) Exponential Growth
t	(price in $1000 units)	(price in $1000 units)
0	50	50
10	75	70.71
20	100	100
30	125	141.42
40	150	200

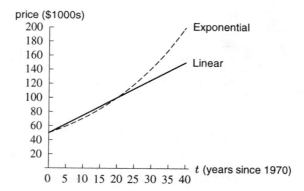

(d)   Since economic growth (inflation, investments) are usually measured in percentage change per year, the exponential model is probably more realistic.

25. (a) Since the growth is exponential, the equation is of the form $P(t) = P_0 a^t$, where $P_0 = 5.6$, the population at time
$t = 0$. Since the growth rate is 1.2%, $a = 1.012$. Thus, $P(t) = 5.6(1.012)^t$, in billions.

   (b)

$$\text{Average rate of change} = \frac{P(2000) - P(1994)}{2000 - 1994}$$
$$= \frac{6.015 - 5.6}{6}$$
$$= 0.069 \text{ billion people per year increase.}$$

   (c)

$$\text{Average rate of change} = \frac{P(2020) - P(2010)}{2020 - 2010}$$
$$= \frac{7.636 - 6.778}{10}$$
$$= 0.086 \text{ billion people per year increase.}$$

Solutions for Section 1.7

1. Since $e^{0.08} = 1.0833$, and $e^{-0.3} = 0.741$, we have

$$P = e^{0.08t} = \left(e^{0.08}\right)^t = (1.0833)^t \text{ and}$$
$$Q = e^{-0.3t} = \left(e^{-0.3}\right)^t = (0.741)^t.$$

5. (a) If the interest is added only once a year (i.e. annually), then at the end of a year we have $1.055x$ where x is the
amount we had at the beginning of the year. After two years, we'll have $1.055(1.055x)$ and after eight years, we'll
have $(1.055)^8 x$. Since we started with \$1000, after eight years we'll have $(1.055)^8(1000) \approx \$1534.69$.

   (b) If an initial deposit of \$$P_0$ is compounded continuously at interest rate r then the account will have $P = P_0 e^{rt}$
after t years. In this case, $P = P_0 e^{rt} = 1000e^{(0.055)(8)} \approx \$1552.71$.

9. In both cases the initial deposit was \$20. Compounding continuously earns more interest than compounding annually at
the same nominal rate. Therefore, curve A corresponds to the account which compounds interest continuously and curve
B corresponds to the account which compounds interest annually. We know that this is the case because curve A is higher
than curve B over the interval, implying that bank account A is growing faster, and thus is earning more money over the
same time period.

13. (a)

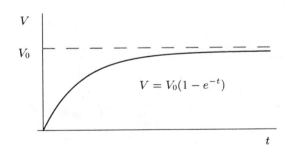

   (b) V_0 represents the terminal velocity of the raindrop or the maximum speed it can attain as it falls (although a raindrop
starting at rest will never quite reach V_0 exactly).

Solutions for Section 1.8

1. $y = \ln e^x$ is a straight line with slope 1, passing through the origin. This is so because $y = \ln e^x = x \ln e = x \cdot 1 = x$. So this function is really $y = x$ in disguise. See Figure 1.10.

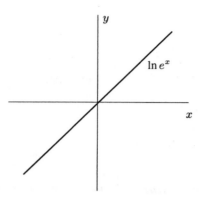

Figure 1.10

5. Taking natural logs of both sides we get
    $$\ln 2 = \ln(1.02^t).$$
    This gives
    $$t \ln 1.02 = \ln 2$$
    or in other words
    $$t = \frac{\ln 2}{\ln 1.02} \approx 35.003.$$

9. Taking natural logs of both sides (and assuming $b \neq 1$) gives
    $$\ln(b^t) = \ln a.$$
    This gives
    $$t \ln b = \ln a$$
    or in other words
    $$t = \frac{\ln a}{\ln b}.$$

13. Dividing both sides by 6 gives
    $$e^{0.5t} = \frac{10}{6} = \frac{5}{3}.$$
    Taking natural logs of both sides we get
    $$\ln(e^{0.5t}) = \ln\left(\frac{5}{3}\right).$$
    This gives
    $$0.5t = \ln\left(\frac{5}{3}\right) = \ln 5 - \ln 3$$
    or in other words
    $$t = 2(\ln 5 - \ln 3) \approx 1.0217.$$

17. Taking natural logs of both sides we get
    $$\ln(5e^{3t}) = \ln(8e^{2t}).$$
    This gives
    $$\ln 5 + \ln(e^{3t}) = \ln 8 + \ln(e^{2t})$$

or in other words

$$\ln(e^{3t}) - \ln(e^{2t}) = \ln 8 - \ln 5.$$

This gives

$$3t - 2t = \ln 8 - \ln 5$$

or in other words

$$t = \ln 8 - \ln 5 \approx 0.47.$$

21. $P = 79(e^{-2.5})^t = 79(0.0821)^t$. Exponential decay because $-2.5 < 0$ or $0.0821 < 1$.

25.   We want $2^t = e^{kt}$ so $2 = e^k$ and $k = \ln 2 = 0.693$. Thus $P = P_0 e^{0.693t}$.

29.   To solve for the value of t

$$100{,}000 = 50{,}000 e^{0.045t},$$

we divide by 50,000 giving

$$2 = e^{0.045t}.$$

Taking natural logs gives

$$\ln 2 = 0.045t$$
$$t = \frac{\ln 2}{0.045} = 15.4 \text{ years}$$

33.   We know that the formula for the balance in an account after t years is

$$P(t) = P_0 e^{rt}$$

where P_0 is the initial deposit and r is the nominal rate. In our case the initial deposit is \$12,000 that is

$$P_0 = 12{,}000$$

and the nominal rate is

$$r = 0.08.$$

Thus we get

$$P(t) = 12{,}000 e^{0.08t}.$$

We are asked to find t such that $P(t) = 20{,}000$, that is we are asked to solve

$$20{,}000 = 12{,}000 e^{0.08t}$$
$$e^{0.08t} = \frac{20{,}000}{12{,}000} = \frac{5}{3}$$
$$\ln e^{0.08t} = \ln(5/3)$$
$$0.08t = \ln 5 - \ln 3$$
$$t = \frac{\ln 5 - \ln 3}{0.08} \approx 6.39$$

Thus after roughly 6.39 years there will be \$20,000 in the account.

37.   Let n be the infant mortality of Senegal. As a function of time t, n is given by

$$n = n_0(0.90)^t.$$

To find when $n = 0.50 n_0$ (so the number of cases has been reduced by 50%), we solve

$$0.50 = (0.90)^t,$$
$$\ln(0.50) = t \ln(0.90),$$
$$t = \frac{\ln(0.50)}{\ln(0.90)} \approx 6.58 \text{ years.}$$

41. Marine catch, M (in millions), is increasing exponentially, so

$$M = M_0 e^{kt}.$$

If we let t be the number of years since 1950, we have $M_0 = 17$ and

$$M = 17 e^{kt}.$$

Since $M = 91$ when $t = 45$, we can solve for k:

$$91 = 17 e^{k(45)}$$
$$91/17 = e^{45k}$$
$$\ln(91/17) = 45k$$
$$k = \frac{\ln(91/17)}{45} = 0.037.$$

Marine catch has increased by 3.7% per year. Since

$$M = 17 e^{0.037t}$$

in the year 2000, we have $t = 50$ and

$$M = 17 e^{0.037(50)} = 108 \text{ million tons.}$$

Solutions for Section 1.9

1. We know that the formula for the total process is

$$P(t) = P_0 a^t$$

where P_0 is the initial size of the process and $a - 1$ is the annual rate. That is,

$$a = 1 + 0.07 = 1.07.$$

We are asked to find the doubling time. Thus, we are asked to solve

$$2P_0 = P_0(1.07)^t$$
$$1.07^t = 2$$
$$\ln 1.07^t = \ln 2$$
$$t \ln 1.07 = \ln 2$$
$$t = \frac{\ln 2}{\ln 1.07} \approx 10.24.$$

Thus, the doubling time is roughly 10.24 years.

Alternatively we could have used the "rule of 70" to get that the doubling time is

$$\frac{70}{7} = 10.$$

5. Since the factor by which the prices have increased after time t is given by $(1.05)^t$, the time after which the prices have doubled solves

$$2 = (1.05)^t$$
$$\ln 2 = \ln(1.05^t) = t \ln(1.05)$$
$$t = \frac{\ln 2}{\ln 1.05} \approx 14.21 \text{ years.}$$

9. We use $P = P_0 e^{kt}$. Since 70% remains after 20 hours, we have $P = 0.70P_0$ when $t = 20$. Solving for k gives:

$$0.70P_0 = P_0 e^{k(20)}$$
$$0.70 = e^{20k}$$
$$\ln(0.70) = 20k$$
$$k = \frac{\ln(0.70)}{20} \approx -0.018.$$

We have $P = P_0 e^{-0.018t}$, and we find t when $P = 0.5P_0$:

$$0.5P_0 = P_0 e^{-0.018t}$$
$$0.5 = e^{-0.018t}$$
$$\ln(0.5) = -0.018t$$
$$t = \frac{\ln(0.5)}{-0.018} \approx 38.5.$$

The half-life is about 38.5 hours.

13. (a) We want to find t such that
$$0.15Q_0 = Q_0 e^{-0.000121t},$$

so $0.15 = e^{-0.000121t}$, meaning that $\ln 0.15 = -0.000121t$, or $t = \dfrac{\ln 0.15}{-0.000121} \approx 15{,}678.7$ years.

(b) Let T be the half-life of carbon-14. Then

$$0.5Q_0 = Q_0 e^{-0.000121T},$$

so $0.5 = e^{-0.000121T}$, or $T = \frac{\ln 0.5}{-0.000121} \approx 5{,}728.5$ years.

17. (a) The total present value for each of the two choices are in the following table. Choice 2 is the preferred choice since it has the larger present value.

Choice 1			Choice 2		
Year	Payment	Present value	Payment	Present value	
0	2000	2000	3000	3000	
1	3000	$3000/(1.05) = 2857.14$	3000	$3000/(1.05) = 2857.14$	
2	4000	$4000/(1.05)^2 = 3628.12$	3000	$3000/(1.05)^2 = 2721.09$	
	Total	8485.26	Total	8578.23	

(b) The difference between the choices is an extra $1000 now ($3000 in Choice 2 instead of $2000 in Choice 1) versus an extra $1000 in two years ($4000 in Choice 1 instead of $3000 in Choice 2). Thus, Choice 2 will have the larger present value no matter what the interest rate.

21. The following table contains the present value of each of the expenses. Since the total present value of the repairs, $255.15, is more than the cost of the service contract, you should buy the service contract.

Present value of repairs		
Year	Repairs	Present Value
1	50	$50/(1.0725) = 46.62$
2	100	$100/(1.0725)^2 = 86.94$
3	150	$150/(1.0725)^3 = 121.59$
	Total	255.15

Solutions for Section 1.10

1.

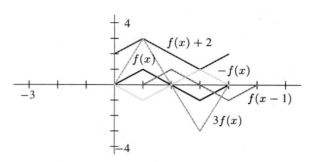

Figure 1.11

5.   (a)   The equation is $y = 2x^2 + 1$. Note that its graph is narrower than the graph of $y = x^2$ which appears in grey.

(b) $y = 2(x^2 + 1)$ moves the graph up one unit and *then* stretches it by a factor of two.

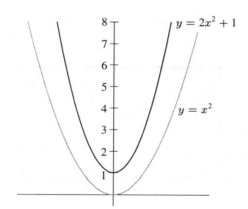

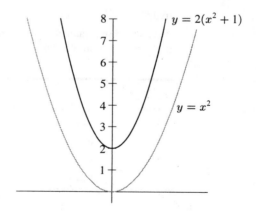

No, the graphs are not the same. Note that stretching vertically leaves any point whose y-value is zero in the same place but moves any other point. This is the source of the difference because if you stretch it first, its lowest point stays at the origin. Then you shift it up by one and its lowest point is $(0, 1)$. Alternatively, if you shift it first, its lowest point is $(0, 1)$ which, when stretched by 2, becomes $(0, 2)$.

(θ)

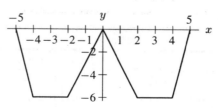

Figure 1.12: Graph of $y = -3f(x)$

13. $f(g(1)) = f(2) \approx 0.4$.

17.

(a) $2H(x)$

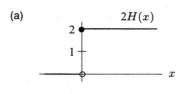

(b) $H(x) + 1$

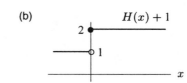

(c) $H(x + 1)$

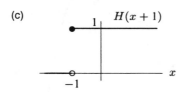

(d)

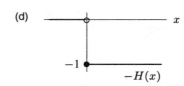
$-H(x)$

(e) $H(-x)$

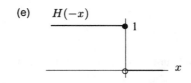

21.   Notice that $f(2) = 4$ and $g(2) = 5$.

(a) $f(2) + g(2) = 4 + 5 = 9$
(b) $f(2) \cdot g(2) = 4 \cdot 5 = 20$
(c) $f(g(2)) = f(5) = 5^2 = 25$
(d) $g(f(2)) = g(4) = 3(4) - 1 = 11$.

25. $\ln(\ln(x))$ means take the ln of the value of the function $\ln x$. (See Figure 1.14.) On the other hand, $\ln^2(x)$ means take the function $\ln x$ and square it. (See Figure 1.15.) For example, consider each of these functions evaluated at e. Since $\ln e = 1$, $\ln^2 e = 1^2 = 1$, but $\ln(\ln(e)) = \ln(1) = 0$. See the graphs in Figures 1.13–1.15. (Note that $\ln(\ln(x))$ is only defined for $x > 1$.)

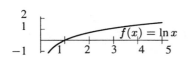

Figure 1.13

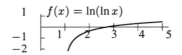

Figure 1.14

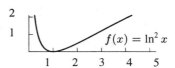

Figure 1.15

Solutions for Section 1.11

1.   (I) Degree ≥ 3, leading coefficient negative.
(II) Degree ≥ 4, leading coefficient positive.
(III) Degree ≥ 4, leading coefficient negative.
(IV) Degree ≥ 5, leading coefficient negative.
(V) Degree ≥ 5, leading coefficient positive.

5.   (a)   The degree of $17 + 8x - 2x^3$ is 3 and the leading coefficient is negative.

(b)   In a large window, the function looks like $-2x^3$. See Figure 1.16.

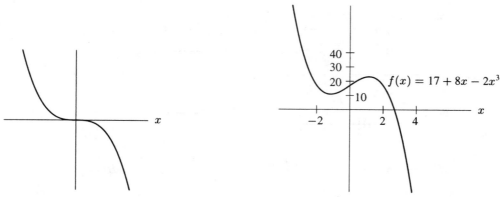

Figure 1.16   **Figure 1.17**

(c)   As $x \to -\infty$, $f(x) \to \infty$. As $x \to \infty$, $f(x) \to -\infty$.
(d)   See Figure 1.17. The function has 2 turning points. Cubics all have 0, 1, or 2 turning points.

9.   (a)   The function $f(x) = 0.2x^7 + 1.5x^4 - 3x^3 + 9x - 15$ has degree 7 and a positive leading coefficient.
      (b)   In a large window, the function looks like $0.2x^7$. See Figure 1.18.

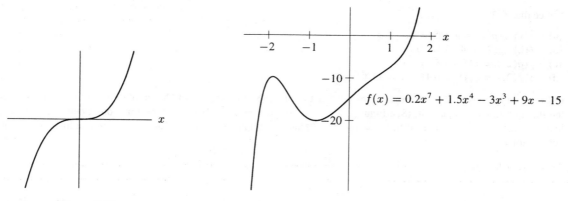

Figure 1.18   **Figure 1.19**

(c)   As $x \to -\infty$, $f(x) \to -\infty$. As $x \to \infty$, $f(x) \to \infty$.
(d)   See Figure 1.19. The function has 2 turning points which is less than the degree.

13.   (a)   We know that

$$q = 3000 - 20p,$$
$$C = 10{,}000 + 35q,$$

and

$$R = pq.$$

Thus, we get

$$C = 10{,}000 + 35q$$
$$= 10{,}000 + 35(3000 - 20p)$$
$$= 10{,}000 + 105{,}000 - 700p$$
$$= 115{,}000 - 700p$$

and

$$R = pq$$
$$= p(3000 - 20p)$$
$$= 3000p - 20p^2.$$

(b)   The graph of the cost and revenue functions are shown in Figure 1.20.

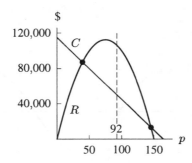

Figure 1.20

(c)   It makes sense that the revenue function looks like a parabola, since if the price is too low, although a lot of people will buy the product, each unit will sell for a pittance and thus the total revenue will be low. Similarly, if the price is too high, although each sale will generate a lot of money, few units will be sold because consumers will be repelled by the high prices. In this case as well, total revenue will be low. Thus, it makes sense that the graph should rise to a maximum profit level and then drop.

(d)   The club makes a profit whenever the revenue curve is above the cost curve in Figure 1.20. Thus, the club makes a profit when it charges roughly between \$40 and \$145.

(e)   We know that the maximal profit occurs when the difference between the revenue and the cost is the greatest. Looking at Figure 1.20, we see that this occurs when the club charges roughly \$92.

Solutions for Section 1.12

1.   It makes sense that sunscreen sales would be lowest in the winter (say from December through February) and highest in the summer (say from June to August). It also makes sense that sunscreen sales would depend almost completely on time of year, so the function should be periodic with a period of 1 year.

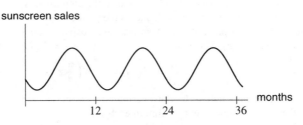

Figure 1.21

5.

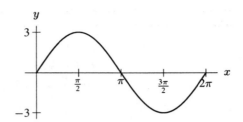

Figure 1.22

The amplitude is 3; the period is 2π.

9.

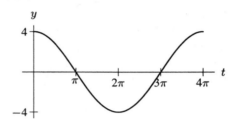

Figure 1.23

The amplitude is 4; the period is 4π.

13.   This graph is an inverted sine curve with amplitude 4 and period π, so it is given by $f(x) = -4\sin(2x)$.

17.   This graph has period 6, amplitude 5 and no vertical or horizontal shift, so it is given by

$$f(x) = 5\sin\left(\frac{2\pi}{6}x\right) = 5\sin\left(\frac{\pi}{3}x\right).$$

21.

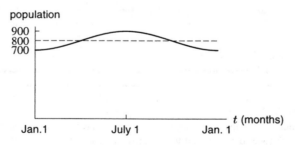

Figure 1.24

25.   (a)   The function appears to vary between 5 and -5, and so the amplitude is 5.
      (b)   The function begins to repeat itself at $x = 8$, and so the period is 8.
      (c)   The function is at its highest point at $x = 0$, so we use a cosine function. It is centered at 0 and has amplitude 5, so we have $f(x) = 5\cos(Bx)$. Since the period is 8, we have $8 = 2\pi/B$ and $B = \pi/4$. The formula is

$$f(x) = 5\cos\left(\frac{\pi}{4}x\right).$$

29.   (a)   Aside from the fact that the graph *looks* like a periodic function, the graph pretty clearly repeats itself and reaches approximately the same minimum and maximum each year.
      (b)   The maximum occurs in the 2nd quarter and the minimum occurs in the 4th quarter. It seems reasonable that people would drink more beer in the summer and less in the winter. Thus, production would be highest the quarter just before summer and lowest the quarter just before winter.
      (c)   The period is 4 quarters or 1 year.

$$\text{Amplitude} = \frac{\text{max} - \text{min}}{2} \approx \frac{55 - 45}{2} = 5 \text{ million barrels}$$

Solutions for Chapter 1 Review

1. (a) The statement $f(12) = 60$ says that when $p = 12$, we have $q = 60$. When the price is \$12, we expect to sell 60 units.

(b) Decreasing, because as price increases, we expect less to be sold.

5. (a) As x gets larger and larger, the value of the function gets closer and closer to 3.

(b) Many answers are possible, such as the following.

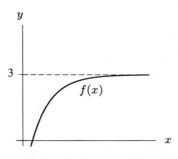

Figure 1.25: Answer to (i)

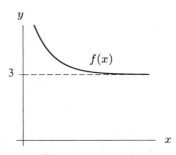

Figure 1.26: Answer to (ii)

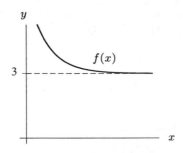

Figure 1.27: Answer to (iii)

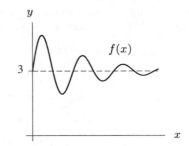

Figure 1.28: Answer to (iv)

9. (a) It was decreasing from March 2 to March 5 and increasing from March 5 to March 9.

(b) From March 5 to 8, the average temperature increased, but the rate of increase went down, from $12°$ between March 5 and 6 to $4°$ between March 6 and 7 to $2°$ between March 7 and 8.

From March 7 to 9, the average temperature increased, and the rate of increase went up, from $2°$ between March 7 and 8 to $9°$ between March 8 and 9.

13. We know that a function is linear if its slope is constant. A function is concave up if the slope increases as t gets larger. And a function is concave down if the slope decreases as t gets larger. Looking at the points $t = 10, t = 20$ and $t = 30$ we see that the slope of $F(t)$ decreases since

$$\frac{F(20) - F(10)}{20 - 10} = \frac{22 - 15}{10} = 0.7$$

while

$$\frac{F(30) - F(20)}{30 - 20} = \frac{28 - 22}{10} = 0.6$$

Looking at the points $t = 10, t = 20$ and $t = 30$ we see that the slope of $G(t)$ is constant

$$\frac{G(20) - G(10)}{20 - 10} = \frac{18 - 15}{10} = 0.3$$

and

$$\frac{G(30) - G(20)}{30 - 20} = \frac{21 - 18}{10} = 0.3$$

Also note that the slope of $G(t)$ is constant everywhere.

Looking at the points $t = 10$, $t = 20$ and $t = 30$ we see that the slope of $H(t)$ increases since

$$\frac{H(20) - H(10)}{20 - 10} = \frac{17 - 15}{10} = 0.2$$

while

$$\frac{H(30) - H(20)}{30 - 20} = \frac{20 - 17}{10} = 0.3$$

Thus $F(t)$ is concave down, $G(t)$ is linear and $H(t)$ is concave up.

17.   (a)

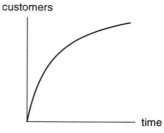

(b)   "The rate at which new people try it" is the rate of change of the total number of people who have tried the product. Thus the statement of the problem is telling you that the graph is concave down—the slope is positive but decreasing, as the graph shows.

21.   By plotting b against $G^{-1/3}$ we see a straight line with slope ≈ 42.5. Alternatively, by calculating $b/G^{-1/3}$ for each of the data points, we find a common value of approximately 42.4. These give an estimate of $a \approx 42.45$. Figure 1.29 shows the function $b = 42.45G^{-1/3}$, which seems to model well the data provided in the problem.

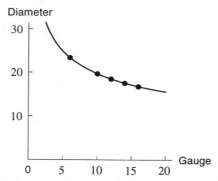

Figure 1.29: The gauge and diameter of a
shotgun's bore and the function
$b = 42.45G^{-1/3}$

25.   (a)   A graph of P against t is shown in Figure 1.30.

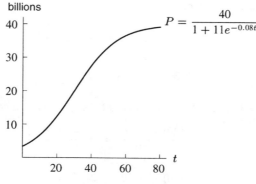

Figure 1.30

(b) We are asked to find the time t such that $P(t) = 20$. We can find this either by tracing along the graph, which tells us that $P = 20$ when $t \approx 30$. Alternatively, we can solve analytically:

$$20 = P(t) = \frac{40}{1 + 11e^{-0.08t}}$$

$$1 + 11e^{-0.08t} = \frac{40}{20} = 2$$

$$11e^{-0.08t} = 1$$

$$e^{-0.08t} = \frac{1}{11}$$

$$\ln e^{-0.08t} = \ln \frac{1}{11}$$

$$-0.08t \approx -2.4$$

$$t \approx \frac{-2.4}{-0.08} = 30.$$

Thus, 30 years from 1990 the population of the world should be 20 billion. In other words the population of the world will be 20 billion in the year 2020.

We are asked to find the time t such that $P(t) = 39.9$. By tracing along the curve, we find $P = 39.9$ when $t \approx 105$. Alternatively, we can solve analytically:

$$39.9 = P(t) = \frac{40}{1 + 11e^{-0.08t}}$$

$$1 + 11e^{-0.08t} = \frac{40}{39.9} = 1.00251$$

$$11e^{-0.08t} = 0.00251$$

$$e^{-0.08t} = \frac{0.00251}{11}$$

$$\ln e^{-0.08t} = \ln \frac{0.00251}{11}$$

$$-0.08t \approx -8.39$$

$$t \approx \frac{-8.39}{-0.08} \approx 105.$$

Thus, 105 years from 1990 the population of the world should be 39.9 billion. In other words the population of the world will be 39.9 billion in the year 2095.

(c) We are asked for the difference in populations at the years 2000 and 1990. That is we are asked for

$$P(10) - P(0).$$

Substituting $t = 0$, we get

$$P(0) = \frac{40}{1 + 11e^{-0.08(0)}}$$

$$= \frac{40}{1 + 11e^0}$$

$$= \frac{40}{1 + 11(1)}$$

$$= \frac{40}{1 + 11}$$

$$= \frac{40}{12}$$

$$\approx 3.33.$$

Thus, in the year 1990 the population would be approximately 3.33 billion. Substituting $t = 10$, we get

$$P(10) = \frac{40}{1 + 11e^{-0.08(10)}}$$

$$= \frac{40}{1 + 11e^{-0.8}}$$

$$= \frac{40}{1 + 11(0.449)}$$

$$= \frac{40}{1 + 4.939}$$

$$= \frac{40}{5.939}$$

$$\approx 6.73.$$

Thus, in the year 2000 the population will be about 6.73 billion. The increase in population between 1990 and 2000 is

$$P(10) - P(0) \approx 6.73 \text{ billion} - 3.33 \text{ billion} = 3.40 \text{ billion}.$$

29. The formula which models compounding interest continuously for t years is $P_0 e^{rt}$ where P_0 is the initial deposit and r is the interest rate. So if we want to double our money while getting 6% interest, we want to solve the following for t:

$$P_0 e^{0.06t} = 2P_0$$

$$e^{0.06t} = 2$$

$$0.06t = \ln(2)$$

$$t = \frac{\ln 2}{0.06} \approx 11.6 \text{ years}$$

Alternatively, by the Rule of Seventy, we have

$$\text{Double time} = \frac{70}{6} \approx 11.67 \text{ years}.$$

33. (a) The graph of the profit is shown in Fig 1.31.

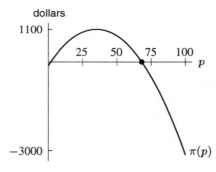

Figure 1.31

Looking at the graph, we can see that the value of p that will maximize profits occurs at about $p = 35$ dollars.

(b) We know that the company will make a profit whenever the profit is greater than zero, or in other words, for all prices p such that the graph of $\pi(p)$ is above the x-axis. Looking at the graph we see that this occurs approximately when the price per skateboard is between $2 and $68.

37. We use the equation $B = Pe^{rt}$. We want to have a balance of $B = \$20,000$ in $t = 6$ years, with an annual interest rate of 10%.

$$20{,}000 = Pe^{(0.1)6}$$

$$P = \frac{20{,}000}{e^{0.6}}$$

$$\approx \$10{,}976.23.$$

41. In effect, your friend is offering to give you $17,000 now in return for the $19,000 lottery payment one year from now. Since $19000/17000 = 1.11764 \cdots$, your friend is charging you 11.7% interest per year, compounded annually. You can expect to get more by taking out a loan as long as the interest rate is less than 11.7%. In particular, if you take out a loan, you have the first lottery check of $19,000 plus the amount you can borrow to be paid back by a single payment of $19000 at the end of the year. At 8.25% interest, compounded annually, the present value of 19,000 one year from now is $19000/(1.0825) = 17551.96$. Therefore the amount you can borrow is the total of the first lottery payment and the loan amount, that is, $19000 + 17551.96 = 36551.96$. So you do better by taking out a one-year loan at 8.25% per year, compounded annually, than by accepting your friend's offer.

45. (a) This moves the graph one unit to the left.

(b) A non-constant polynomial tends toward $+\infty$ or $-\infty$ as $x \to \infty$. This polynomial p does not. Therefore, p must be a constant function, i.e. its graph is a horizontal line.

49. (a) The rate R is the difference of the rate at which the glucose is being injected, which is given to be constant, and the rate at which the glucose is being broken down, which is given to be proportional to the amount of glucose present. Thus we have the formula

$$R = k - aG$$

where k is the rate that the glucose is being injected, a is the constant relating the rate that it is broken down to the amount present, and G is the amount present.

(b)

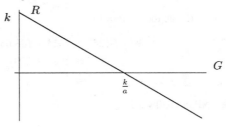

53. The graph looks like a sine function of amplitude 7 and period 10. If the equation is of the form

$$y = 7\sin(kt),$$

then $2\pi/k = 10$, so $k = \pi/5$. Therefore, the equation is

$$y = 7\sin\left(\frac{\pi t}{5}\right).$$

Solutions to Problems on Fitting Formulas to Data

1. (a) A line seems to fit the data in Figure 1.32 reasonably well.

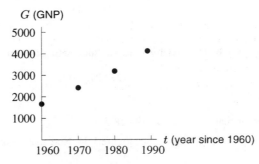

Figure 1.32: GNP as a function of year

(b) If we let G represent the GNP and t represent the years that have passed since 1960, we get from a calculator the linear regression equation:

$$G = F(t) = 1623 + 83.65t.$$

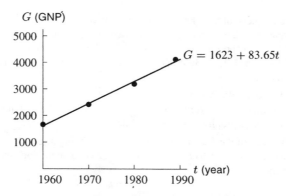

Figure 1.33: GNP data and linear regression function

(c)   Using our equation, we estimate the GNP in 1985 as

$$F(25) = 1623 + (83.65)(25)$$
$$\approx 1623 + 2091$$
$$= 3714.$$

We can estimate the GNP in 2020 with

$$F(60) = 1623 + (83.65)(60) = 1623 + 5019$$
$$= 6642.$$

Our estimate is likely to work better for the year 1985 than for the year 2020. This is because this sort of estimation works better for years that are closer to the existing data points. Thus, our estimation will be more accurate for 1985, which is within the range of the data points given and hence involves interpolation, than 2020, for which we need extrapolation.

5.   (a)   The annual growth rate of an exponential function of the form Ar^t is just $(r - 1)$, analogous to the rate of interest in interest problems. Thus, in the given function

$$r - 1 = 1.0026 - 1.00$$
$$= 0.0026.$$

This means that the CO_2 concentration grows by 0.26% every year.

(b)   The CO_2 concentration given by the model for 1900, when we substitute $t = 0$ into the function, is

$$C = 272.27(1.0026)^0$$
$$= 272.27(1)$$
$$= 272.27.$$

The CO_2 concentration given by the model for 1980 is, substituting $t = 80$

$$C = 272.27(1.0026)^{80}$$
$$\approx 335.1.$$

This is not too far from the real concentration of 338.5.

9.   (a)   A linear model seems to provide the best fit. See Figure 1.34. Note that we are looking at a data set with a negative slope.

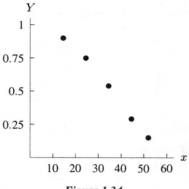

Figure 1.34

(b)   The best linear regression function for this data set is

$$Y = -0.021x + 1.23$$

The slope of this function is -0.021. The negative sign indicates that the farther the distance from the goal line, the less likely a field goal kick is to succeed. The success rate goes down by about 0.021 for each additional yard from the goal line. See Figure 1.35.

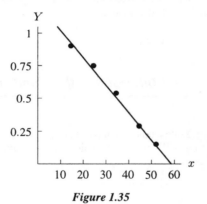

Figure 1.35

(c)   The best exponential regression function is

$$Y = 2.17(0.954)^x,$$

which is plotted in Figure 1.36.

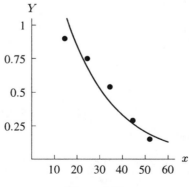

Figure 1.36

We get the success rate predicted by substituting $x = 50$ into the exponential regression function

$$Y = 2.17(0.954)^x$$
$$= 2.17(0.954)^{50}$$
$$\approx 2.17(0.095)$$
$$\approx 0.206.$$

Thus, the predicted success rate is 20.6%.

(d)   Of the two graphs, the linear model seems to fit the data best.

13.   (a)

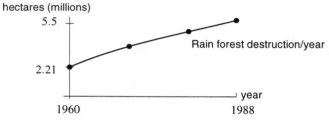

hectares (millions)

Rain forest destruction/year

Figure 1.37

(b)   The data are increasing and the curve is slightly concave down. The fact that the data are increasing means that more hectares of rain forest are being destroyed every year. The fact that the curve is concave down means the rate of increase of rain forest destruction is decreasing.

(c)   Using a graphing calculator, we get $y \approx -1885 + 249 \ln x$.

(d)   When $x = 2000$, $y \approx 7.6 \times 10^6$ hectares.

Solutions to Problems on Compound Interest and the Number e

1.   For $20 \le x \le 100$, $0 \le y \le 1.2$, this function looks like a horizontal line at $y \approx 1.0725$ (In fact, the graph approaches this line from below.) Now, $e^{0.07} \approx 1.0725$, which strongly suggests that, as we already know, as $x \to \infty$, $\left(1 + \dfrac{0.07}{x}\right)^x \to e^{0.07}$.

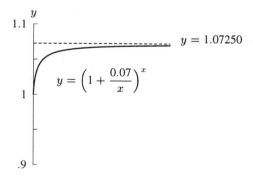

5.   (a)   (i)

$$\left(1 + \frac{0.05}{1000}\right)^{1000} = 1.05126978\ldots,$$

so the effective annual yield is $5.126978\ldots\%$.

(ii)

$$\left(1 + \frac{0.05}{10000}\right)^{10000} = 1.05127096\ldots,$$

so the effective annual yield is $5.127096\ldots\%$.

(iii)

$$\left(1 + \frac{0.05}{100000}\right)^{100000} = 1.05127108\ldots,$$

so the effective annual yield is $5.127108\ldots\%$.

(b)   The effective annual rates in part (a) are closing in on 5.127%, so this is the effective annual yield for a 5% annual rate compounded continuously.

(c) $e^{0.05} = 1.05127109\ldots$. Since continuous compounding is equivalent to multiplying by $e^{0.05}$, the effective annual yield for continuous compounding is $0.05127109\ldots \approx 5.127\%$.

9.   We know that for a given annual rate, the higher the frequency of compounding, the higher the effective annual yield. So the effective yield of (a) will be greater than that of (c) which is greater than that of (b). Also, the effective annual yield of (e) will be greater than that of (d). Now the effective annual yield of (e) will be less than the effective annual yield of 5.5% annual rate, compounded twice a year, and the latter will be less than the yield from (b). Thus $d < e < b < c < a$. Matching these up with our choices, we get

(d) I, (e) II, (b) III, (c) IV, (a) V.

CHAPTER TWO

Solutions for Section 2.1

1.

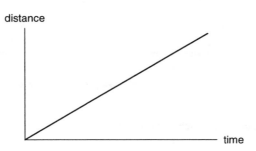

distance

time

5. (a) The average velocity between $t = 3$ and $t = 5$ is

$$\frac{\text{Distance}}{\text{Time}} = \frac{s(5) - s(3)}{5 - 3} = \frac{25 - 9}{2} = \frac{16}{2} = 8 \text{ ft/sec}.$$

(b) Using an interval of size 0.1, we have

$$\left(\begin{array}{c} \text{Instantaneous velocity} \\ \text{at } t = 3 \end{array}\right) \approx \frac{s(3.1) - s(3)}{3.1 - 3} = \frac{9.61 - 9}{0.1} = 6.1.$$

Using an interval of size 0.01, we have

$$\left(\begin{array}{c} \text{Instantaneous velocity} \\ \text{at } t = 3 \end{array}\right) \approx \frac{s(3.01) - s(3)}{3.01 - 3} = \frac{9.0601 - 9}{0.01} = 6.01.$$

From this we guess that the instantaneous velocity at $t = 3$ is about 6 ft/sec.

9. For the interval $0 \le t \le 0.8$, we have

$$\left(\begin{array}{c} \text{Average velocity} \\ 0 \le t \le 0.8 \end{array}\right) = \frac{s(0.8) - s(0)}{0.8 - 0} = \frac{6.5}{0.8} = 8.125 \text{ ft/sec}.$$

$$\left(\begin{array}{c} \text{Average velocity} \\ 0 \le t \le 0.2 \end{array}\right) = \frac{s(0.2) - s(0)}{0.2 - 0} = \frac{0.5}{0.2} = 2.5 \text{ ft/sec}.$$

$$\left(\begin{array}{c} \text{Average velocity} \\ 0.2 \le t \le 0.4 \end{array}\right) = \frac{s(0.4) - s(0.2)}{0.4 - 0.2} = \frac{1.3}{0.2} = 6.5 \text{ ft/sec}.$$

To find the velocity at $t = 0.2$, we find the average velocity to the right of $t = 0.2$ and to the left of $t = 0.2$ and average them. So a reasonable estimate of the velocity at $t = 0.2$ is the average of $\frac{1}{2}(6.5 + 2.5) = 4.5$ ft/sec.

13. (a) When $t = 0$, the ball is on the bridge and its height is $f(0) = 36$, so the bridge is 36 feet above the ground.

(b) After 1 second, the ball's height is $f(1) = -16 + 50 + 36 = 70$ feet, so it traveled $70 - 36 = 34$ feet in 1 second, and its average velocity was 34 ft/sec.

(c) At $t = 1.001$, the ball's height is $f(1.001) = 70.017984$ feet, and its velocity about

$$\text{Velocity} = \frac{70.017984 - 70}{1.001 - 1} = 17.984 \approx 18 \text{ ft/sec}.$$

(d) The graph is shown in Figure 2.1. Since the coordinates of the peak are about $(1.6, 75)$ the ball reaches a height of about 75 feet. The velocity of the ball is zero when it is at its peak since the tangent is horizontal there.

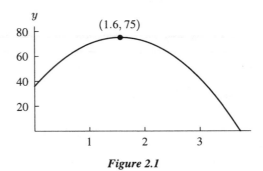

Figure 2.1

(e) The ball reaches its maximum height when $t \approx 1.6$.

Solutions for Section 2.2

1. (a) The average rate of change is the slope of the secant line in Figure 2.2, which shows that this slope is positive.
   (b) The instantaneous rate of change is the slope of the graph at $x = 3$, which we see from Figure 2.3 is negative.

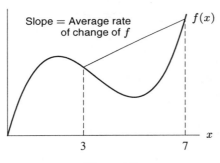

Figure 2.2

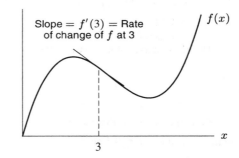

Figure 2.3

5. We know that $f'(2)$ is the derivative of $f(x) = x^3 - 2x$ at 2. This is the same as the rate of change of $x^3 - 2x$ at 2. We estimate this by looking at the average rate of change over intervals near 2. If we use the intervals $1.999 \le x \le 2$ and $2 \le x \le 2.001$, we see that

$$\begin{pmatrix} \text{Average rate of change} \\ \text{on } 1.999 \le x \le 2 \end{pmatrix} = \frac{[2^3 - 2(2)] - [1.999^3 - 2(1.999)]}{2 - 1.999} = \frac{4 - 3.990}{0.001} = 10,$$

$$\begin{pmatrix} \text{Average rate of change} \\ \text{on } 2 \le x \le 2.001 \end{pmatrix} = \frac{[2.001^3 - 2(2.001)] - [2^3 - 2(2)]}{2.001 - 2} = \frac{4.010 - 4}{0.001} = 10.$$

It appears that the rate of change of $f(x)$ at $x = 2$ is 10, so we estimate $f'(2) = 10$.

9. (a) At points A, B, and D the function is increasing. Therefore, the derivative of the function at points A, B, and D is positive. At points C and F, the function is decreasing. Therefore, the derivative of the function at points C and F is negative. At point E, the function is neither decreasing nor increasing. Therefore, the derivative of the function at point E is zero.
   (b) The derivative is the most positive where the graph of the function is the steepest and increasing, as at point D. The derivative is the most negative where the graph of the function is the steepest and decreasing, as at point F.

13.

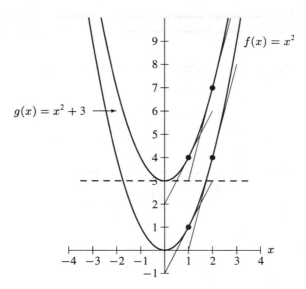

Figure 2.4

(a) The tangent line to the graph of $f(x) = x^2$ at $x = 0$ coincides with the x-axis and therefore is horizontal (slope $= 0$). The tangent line to the graph of $g(x) = x^2 + 3$ at $x = 0$ is the dashed line indicated in the figure and it also has a slope equal to zero. Therefore both tangent lines at $x = 0$ are parallel.

   We see in Figure 2.4 that the tangent lines at $x = 1$ appear parallel, and the tangent lines at $x = 2$ appear parallel. The slopes of the tangent lines at any value $x = a$ will be equal.

(b) Adding a constant shifts the graph vertically, but does not change the slope of the curve.

17. (a) A graph of $f(x)$ and the tangent line to the point $(2, \ln 2)$ are shown in Figure 2.5. Looking at the graph we see that as x goes from 2 to 4 the corresponding y-increment in the tangent line is approximately of length 1, making the slope of the tangent and the derivative at $x = 2$

$$f'(2) \approx \frac{1}{4 - 2} = 0.5$$

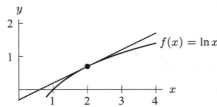

Figure 2.5

(b) We know that

$$f'(2) \approx \frac{f(2 + h) - f(2)}{h} \qquad \text{for small} \quad h$$
$$= \frac{\ln(2 + h) - \ln(2)}{h}$$

Substituting small values of h gives Table 2.1.

TABLE 2.1

h	0.1	0.01	0.001
$f'(2)$	0.4879	0.4988	0.4999

Thus this approximation of the derivative also gives us

$$f'(2) \approx 0.5$$

Solutions for Section 2.3

1. The graph is that of the line $y = -2x + 2$. The slope, and hence the derivative, is -2.

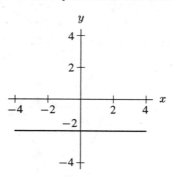

5.

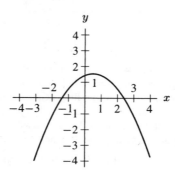

9.

(a)

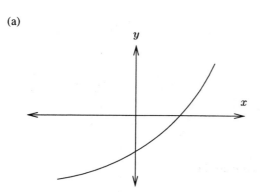

(b)

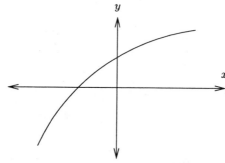

(c)

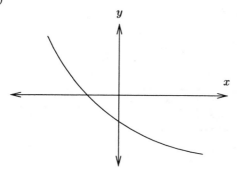

(d)

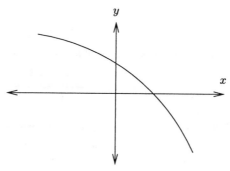

13.

TABLE 2.2

x	$f(x)$	x	$f(x)$	x	$f(x)$
1.998	2.6587	2.998	8.9820	3.998	21.3013
1.999	2.6627	2.999	8.9910	3.999	21.3173
2.000	2.6667	3.000	9.0000	4.000	21.3333
2.001	2.6707	3.001	9.0090	4.001	21.3493
2.002	2.6747	3.002	9.0180	4.002	21.3653

Near 2, the values of $f(x)$ seem to be increasing by 0.004 for each increase of 0.001 in x, so the derivative appears to be $\frac{0.004}{0.001} = 4$. Near 3, the values of $f(x)$ are increasing by 0.009 for each step of 0.001, so the derivative appears to be 9. Near 4, $f(x)$ increases by 0.016 for each step of 0.001, so the derivative appears to be 16. The pattern seems to be, then, that at a point x, the derivative of $f(x) = \frac{1}{3}x^3$ is $f'(x) = x^2$.

17. (a) x_3 (b) x_4 (c) x_5 (d) x_3

21.

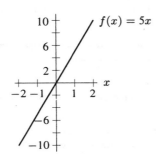
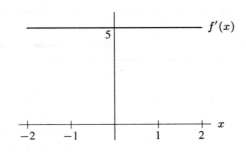

25.

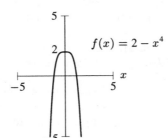
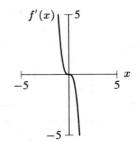

Figure 2.6

29.

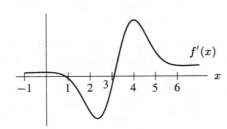

33.   (a)   A possible graph is shown in Figure 2.7

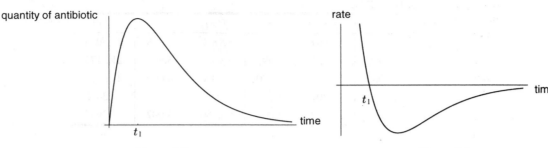

Figure 2.7   **Figure 2.8**

(b)   The injection puts a reservoir of antibiotics in the muscle which begins to diffuse into the blood. As the antibiotic diffuses into the bloodstream, it begins to leave the blood either through normal metabolic action or absorption by some other organ. For a while the antibiotic diffuses into the blood faster than it is lost and its concentration rises, but as the reservoir in the muscle is drawn down, the diffusion rate into the blood decreases and eventually becomes less than the loss rate. After that, the concentration in the blood goes down. But the rate of decrease in concentration gets smaller as we approach the time when all the antibiotic is lost. This is shown in Figure 2.8.

Solutions for Section 2.4

1.   (Note that we are considering the average temperature of the yam, since its temperature is different at different points inside it.)

(a)   It is positive, because the temperature of the yam increases the longer it sits in the oven.

(b)   The units of $f'(20)$ are °F/min. $f'(20) = 2$ means that at time $t = 20$ minutes, the temperature T increases by approximately 2°F for each additional minute in the oven.

5.   Units of $C'(r)$ are dollars/percent. Approximately, $C'(r)$ means the additional amount needed to pay off the loan when the interest rate is increased by 1%. The sign of $C'(r)$ is positive, because increasing the interest rate will increase the amount it costs to pay off a loan.

9.   (a)   This means that investing the $1000 at 5% would yield $1649 after 10 years.

(b)   Writing $g'(r)$ as dB/dt, we see that the units of dB/dt are dollars per percent (interest). We can interpret dB as the extra money earned if interest rate is increased by dr percent. Therefore $g'(5) = \frac{dB}{dr}|_{r=5} \approx 165$ means that the balance, at 5% interest, would increase by about $165 if the interest rate were increased by 1%. In other words, $g(6) \approx g(5) + 165 = 1649 + 165 = 1814$.

13.   (a)   Since t represents the number of days from now, we are told $f(0) = 80$ and $f'(0) = 0.50$.

(b)

$$f(10) \approx \text{value now} + \text{change in value in 10 days}$$
$$= 80 + 0.50(10)$$
$$= 80 + 5$$
$$= 85.$$

In 10 days, we expect that the mutual fund will be worth about $85 a share.

17.   (a)   The company hopes that increased advertising always brings in more customers instead of turning them away. Therefore, it hopes $f'(a)$ is always positive.

(b)   If $f'(100) = 2$, it means that if the advertising budget is $100,000, each extra dollar spent on advertising will bring in $2 worth of sales. If $f'(100) = 0.5$, each dollar above $100 thousand spent on advertising will bring in $0.50 worth of sales.

(c)   If $f'(100) = 2$, then as we saw in part (b), spending slightly more than $100,000 will increase revenue by an amount greater than the additional expense, and thus more should be spent on advertising. If $f'(100) = 0.5$, then the increase in revenue is less than the additional expense, hence too much is being spent on advertising. The optimum amount to spend, a, is an amount that makes $f'(a) = 1$. At this point, the increases in advertising expenditures just pay for themselves. If $f'(a) < 1$, too much is being spent; if $f'(a) > 1$, more should be spent.

21. (a) Clearly the population of the US at any instant is an integer that varies up and down every few seconds as a child is born, a person dies, or a new immigrant arrives. Since these events cannot usually be assigned to an exact instant, the population of the US at any given moment might actually be indeterminate. If we count in units of a thousand, however, the population appears to be a smooth function that has been rounded to the nearest thousand.

Major land acquisitions such as the Louisiana Purchase caused larger jumps in the population, but since the census is taken only every ten years and the territories acquired were rather sparsely populated, we cannot see these jumps in the census data.

(b) We can regard rate of change of the population for a particular time t as representing an estimate of how much the population will increase during the year after time t.

(c) Many economic indicators are treated as smooth, such as the Gross National Product, the Dow Jones Industrial Average, volumes of trading, and the price of commodities like gold. But these figures only change in increments, and not continuously.

Solutions for Section 2.5

1. (a) increasing, concave up
    (b) decreasing, concave down

5. The derivative is positive on those intervals where the function is increasing and negative on those intervals where the function is decreasing. Therefore, the derivative is positive on the intervals $0 < t < 0.4$ and $1.7 < t < 3.4$, and negative on the intervals $0.4 < t < 1.7$ and $3.4 < t < 4$.

The second derivative is positive on those intervals where the graph of the function is concave up and negative on those intervals where the graph of the function is concave down. Therefore, the second derivative is positive on the interval $1 < t < 2.6$ and negative on the intervals $0 < t < 1$ and $2.6 < t < 4$.

9. (a)

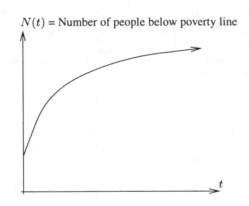

$N(t)$ = Number of people below poverty line

t

(b) dN/dt is positive, since people are still slipping below the poverty line. d^2N/dt^2 is negative, since the rate at which people are slipping below the poverty line, dN/dt, is decreasing.

13. (a) $f'(0.6) \approx \dfrac{f(0.8) - f(0.6)}{0.8 - 0.6} = \dfrac{4.0 - 3.9}{0.2} = 0.5.$ $f'(0.5) \approx \dfrac{f(0.6) - f(0.4)}{0.6 - 0.4} = \dfrac{0.4}{0.2} = 2.$

(b) Using the values of f' from part (a), we get $f''(0.6) \approx \dfrac{f'(0.6) - f'(0.5)}{0.6 - 0.5} = \dfrac{0.5 - 2}{0.2} = \dfrac{-1.5}{0.2} = -7.5.$

(c) The maximum value of f is probably near $x = 0.8$. The minimum value of f is probably near $x = 0.3$.

17. (a) B (where $f', f'' > 0$) and E (where $f', f'' < 0$)
    (b) A (where $f = f' = 0$) and D (where $f' = f'' = 0$)

Solutions for Section 2.6

1. We know $MC \approx C(1,001) - C(1,000)$. Therefore, $C(1,001) \approx C(1,000) + MC$ or $C(1,001) \approx 5000 + 25 = 5025$ dollars.

Since we do not know $MC(999)$, we will assume that $MC(999) = MC(1,000)$. Therefore:

$$MC(999) = C(1,000) - C(999).$$

Then:

$$C(999) \approx C(1,000) - MC(999) = 5,000 - 25 = 4,975 \text{ dollars.}$$

Alternatively, we can reason that

$$MC(1,000) \approx C(1,000) - C(999),$$

so

$$C(999) \approx C(1,000) - MC(1,000) = 4,975 \text{ dollars.}$$

Now for $C(1,000)$, we have

$$C(1,100) \approx C(1,000) + MC \cdot 100.$$

Since $1,100 - 1,000 = 100$,

$$C(1,100) \approx 5,000 + 25 \times 100 = 5,000 + 2,500 = 7,500 \text{ dollars.}$$

5. Marginal cost $= C'(q)$. Therefore, marginal cost at q is the slope of the graph of $C(q)$ at q. We can see that the slope at $q = 5$ is greater than the slope at $q = 30$. Therefore, marginal cost is greater at $q = 5$. We see that, at $q = 20$, the slope is practically zero while at $q = 40$ the slope is positive. Therefore, marginal cost at $q = 40$ is greater than marginal cost at $q = 20$.

9. (a) Since Profit $=$ Revenue $-$ Cost, we can calculate $\pi(q) = R(q) - C(q)$ for each of the q values given:

q	0	100	200	300	400	500
$R(q)$	0	500	1000	1500	2000	2500
$C(q)$	700	900	1000	1100	1300	1900
$\pi(q)$	-700	-400	0	400	700	600

We see that maximum profit is $700 and it occurs when the production level q is 400. See Figure 2.9.

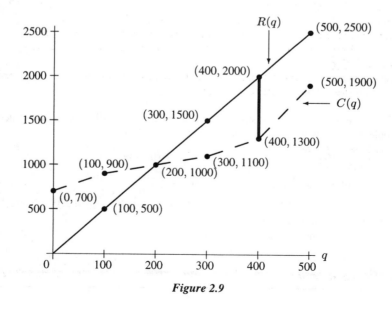

Figure 2.9

(b) Since revenue is $500 when $q = 100$, the selling price is $5 per unit.
(c) Since $C(0) = \$700$, the fixed costs are $700.

13. (a) The profit earned by the 51st is the revenue earned by the 51st item minus the cost of producing the 51st item. This can be approximated by

$$\pi'(50) = R'(50) - C'(50) = 84 - 75 = \$9.$$

Thus the profit earned from the 51st item will be approximately \$9.

(b) The profit earned by the 91st item will be the revenue earned by the 91st item minus the cost of producing the 91st item. This can be approximated by

$$\pi'(90) = R'(90) - C'(90) = 68 - 71 = -\$3.$$

Thus, approximately three dollars are lost in the production of the 91st item.

(c) If $R'(78) > C'(78)$, production of a 79th item would increase profit. If $R'(78) < C'(78)$, production of one less item would increase profit. Since profit is maximized at $q = 78$, we must have

$$C'(78) = R'(78).$$

Solutions for Chapter 2 Review

1.

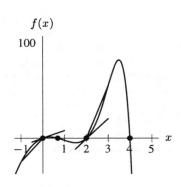

$f(x)$

100

(a) The graph shows four different zeros in the interval, at $x = 0$, $x = 2$, $x = 4$ and $x \approx 0.7$.
(b) At $x = 0$ and $x = 2$, we see that the tangent has a positive slope so f is increasing.
      At $x = 4$, we notice that the tangent to the curve has negative slope, so f is decreasing.
(c) Comparing the slopes of the secant lines at these values, we can see that the average rate of change of f is greater on the interval $2 \leq x \leq 3$.
(d) Looking at the tangents of the function at $x = 0$ and $x = 2$, we see that the slope of the tangent at $x = 2$ is greater. Thus, the instantaneous rate of change of f is greater at $x = 2$.

5. See Figure 2.10.

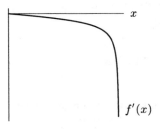

x

$f'(x)$

Figure 2.10

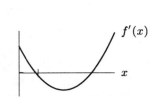

$f'(x)$

x

Figure 2.11

9. See Figure 2.11.

13. Estimating the slope of the lines in Figure 2.12, we find that $f'(-2) \approx 1.0$, $f'(-1) \approx 0.3$, $f'(0) \approx -0.5$, and $f'(2) \approx -1$.

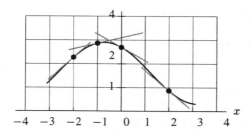

Figure 2.12

17. (a) The function appears to be decreasing and concave down, and so we conjecture that f' is negative and that f'' is negative.

(b) We use difference quotients to the right:
$$f'(2) \approx \tfrac{137-145}{4-2} = -4$$
$$f'(8) \approx \tfrac{56-98}{10-8} = -21.$$

21. (a) The marginal cost at $q = 400$ is the slope of the tangent line to $C(q)$ at $q = 400$. Looking at the graph, we can estimate a slope of about 1. Thus, the marginal cost is $1.

(b) At $q = 500$, we can see that slope of the cost function is greater than the slope of the revenue function. Thus, the marginal cost is greater than the marginal revenue and thus the 500th item will incur a loss. So, the company should not produce the 500th item.

(c) The quantity which maximizes profit is at the point where marginal costs are just about to exceed marginal revenue. This occurs when the slope of $R(q)$ equals $C(q)$, which occurs at $q = 400$. Thus, the company should produce 400 items.

25. (a) Slope of tangent line $\approx \frac{\sqrt{4.001}-\sqrt{4}}{0.001} = 0.249984$. Hence the slope of the tangent line is about 0.25.

(b) The tangent line has slope 0.25 and contains the point $(4, 2)$, so its equation is
$$y - 2 = 0.25(x - 4)$$
$$y - 2 = 0.25x - 1$$
$$y = 0.25x + 1$$

(c) We have $f(x) = kx^2$. If $(4, 2)$ is on the graph of f, then $f(4) = 2$, so $k \cdot 4^2 = 2$. Thus $k = \tfrac{1}{8}$, and $f(x) = \tfrac{1}{8}x^2$.

(d) To find where the graph of f crosses the line $y = 0.25x + 1$, we solve:
$$\frac{1}{8}x^2 = 0.25x + 1$$
$$x^2 = 2x + 8$$
$$x^2 - 2x - 8 = 0$$
$$(x - 4)(x + 2) = 0$$
$$x = 4 \text{ or } x = -2$$
$$f(-2) = \frac{1}{8}(4) = 0.5$$

Therefore, $(-2, 0.5)$ is the other point of intersection. (Of course, $(4, 2)$ is a point of intersection; we know that from the start.)

29. (a) The statement $f(15) = 200$ tells us that when the price is $15, we sell about 200 units of the product.

(b) The statement $f'(15) = -25$ tells us that if we increase the price by $1 (from 15), we will sell about 25 fewer units of the product.

Solutions to Limits and the Definition of the Derivative ━━━

1. The answers are marked in Figure 2.13.

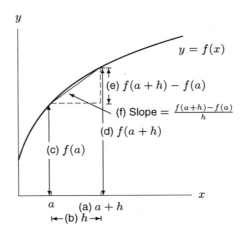

Figure 2.13

5. Using $h = 0.1, 0.01, 0.001$, we see

$$\frac{(3+0.1)^3 - 27}{0.1} = 27.91$$

$$\frac{(3+0.01)^3 - 27}{0.01} = 27.09$$

$$\frac{(3+0.001)^3 - 27}{0.001} = 27.009.$$

These calculations suggest that $\lim\limits_{h \to 0} \dfrac{(3+h)^3 - 27}{h} = 27$.

9. Yes, $f(x)$ is continuous on $0 \le x \le 2$.

13. Yes: $f(x) = x + 2$ is continuous for all values of x.

17. No: $f(x) = 1/(x - 1)$ is not continuous on any interval containing $x = 1$.

21. Since we can't make a fraction of a pair of pants, the number increases in jumps, so the function is not continuous.

25. Using the definition of the derivative, we have

$$f'(x) = \lim_{h \to 0} \frac{f(x+h) - f(x)}{h} = \lim_{h \to 0} \frac{3(x+h)^2 - 3x^2}{h}$$

$$= \lim_{h \to 0} \frac{3(x^2 + 2xh + h^2) - 3x^2}{h}$$

$$= \lim_{h \to 0} \frac{3x^2 + 6xh + 3h^2 - 3x^2}{h}$$

$$= \lim_{h \to 0} \frac{6xh + 3h^2}{h} = \lim_{h \to 0} \frac{h(6x + 3h)}{h}.$$

As h gets very close to zero (but not equal to zero), we can cancel the h in the numerator and denominator to leave the following:

$$f'(x) = \lim_{h \to 0} (6x + 3h).$$

As $h \to 0$, we have $f'(x) = 6x$.

CHAPTER THREE

Solutions for Section 3.1

1. (a) The velocity is 30 miles/hour for the first 2 hours, 40 miles/hour for the next 1/2 hour, and 20 miles/hour for the last 4 hours. The entire trip lasts $2 + 1/2 + 4 = 6.5$ hours, so we need a scale on our horizontal (time) axis running from 0 to 6.5. Between $t = 0$ and $t = 2$, the velocity is constant at 30 miles/hour, so the velocity graph is a horizontal line at 30. Likewise, between $t = 2$ and $t = 2.5$, the velocity graph is a horizontal line at 40, and between $t = 2.5$ and $t = 6.5$, the velocity graph is a horizontal line at 20. The graph is shown in Figure 3.1.

   (b) How can we visualize distance traveled on the velocity graph given in Figure 3.1? The velocity graph looks like the top edges of three rectangles. The distance traveled on the first leg of the journey is (30 miles/hour)(2 hours), which is the height times the width of the first rectangle in the velocity graph. The distance traveled on the first leg of the trip is equal to the area of the first rectangle. Likewise, the distances traveled during the second and third legs of the trip are equal to the areas of the second and third rectangles in the velocity graph. It appears that distance traveled is equal to the area under the velocity graph.

   In Figure 3.2, the area under the velocity graph in Figure 3.1 is shaded. Since this area is three rectangles and the area of each rectangle is given by Height × Width, we have

$$\text{Total area} = (30)(2) + (40)(1/2) + (20)(4)$$
$$= 60 + 20 + 80 = 160.$$

The area under the velocity graph is equal to distance traveled.

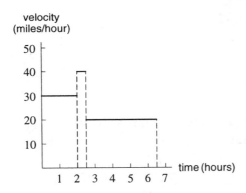

Figure 3.1: Velocity graph

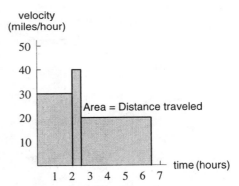

Figure 3.2: The area under the velocity graph gives distance traveled

5. (a) Note that 15 minutes equals 0.25 hours. Lower estimate $= 11(0.25) + 10(0.25) = 5.25$ miles. Upper estimate $= 12(0.25) + 11(0.25) = 5.75$ miles.

   (b) Lower estimate $= 11(0.25) + 10(0.25) + 10(0.25) + 8(0.25) + 7(0.25) + 0(0.25) = 11.5$ miles. Upper estimate $= 12(0.25) + 11(0.25) + 10(0.25) + 10(0.25) + 8(0.25) + 7(0.25) = 14.5$ miles.

9. Just counting the squares (each of which has area 10), and allowing for the broken squares, we can see that the area under the curve from 0 to 6 is between 140 and 150. Hence the distance traveled is between 140 and 150 meters.

Solutions for Section 3.2

1. (a) If $\Delta t = 4$, then $n = 2$. We have:

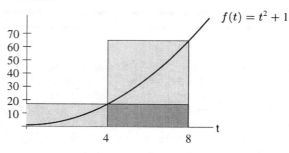

Figure 3.3

Underestimate of total change $= f(0)\Delta t + f(4)\Delta t = (1)(4) + (17)(4) = 4 + 68 = 72$.

Overestimate of total change $= f(4)\Delta t + f(8)\Delta t = (17)(4) + (65)(4) = 68 + 260 = 328$.

(b) If $\Delta t = 2$, then $n = 4$. We have:

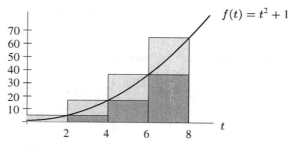

Figure 3.4

Underestimate of total change $= f(0)\Delta t + f(2)\Delta t + f(4)\Delta t + f(6)\Delta t$
$$= (1)(2) + (5)(2) + (17)(2) + (37)(2) = 120.$$

Overestimate of total change $= f(2)\Delta t + f(4)\Delta t + f(6)\Delta t + f(8)\Delta t$
$$= (5)(2) + (17)(2) + (37)(2) + (65)(2) = 248.$$

(c) If $\Delta t = 1$, then $n = 8$.

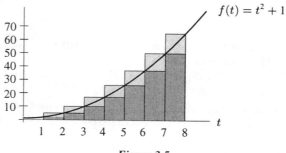

Figure 3.5

Underestimate of total change

$$= f(0)\Delta t + f(1)\Delta t + f(2)\Delta t + f(3)\Delta t + f(4)\Delta t + f(5)\Delta t + f(6)\Delta t + f(7)\Delta t$$
$$= (1)(1) + (2)(1) + (5)(1) + (10)(1) + (17)(1) + (26)(1) + (37)(1) + (50)(1) = 148$$

Overestimate of total change
$$= f(1)\Delta t + f(2)\Delta t + f(3)\Delta t + f(4)\Delta t + f(5)\Delta t + f(6)\Delta t + f(7)\Delta t + f(8)\Delta t$$
$$= (2)(1) + (5)(1) + (10)(1) + (17)(1) + (26)(1) + (37)(1) + (50)(1) + (65)(1) = 212$$

5. (a) See Figure 3.6.

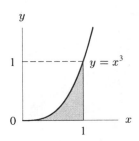

Figure 3.6

$$\int_0^1 x^3\, dx = \text{area shaded, which is less than 0.5. Rough estimate is about 0.3.}$$

(b) $\int_0^1 x^3\, dx = 0.25$

9. $\int_0^5 x^2\, dx = 41.7$

13. $\int_1^2 2^x\, dx = 2.9$

17. $\int_{-3}^3 e^{-t^2}\, dt \approx 2\int_0^3 e^{-t^2}\, dt = 2(0.886) = 1.772$

21. We estimate the integral by finding left- and right-hand sums and averaging them:

$$\text{Left-hand sum} = (100)(4) + (88)(4) + (72)(4) + (50)(4) = 1240,$$

and

$$\text{Right-hand sum} = (88)(4) + (72)(4) + (50)(4) + (28)(4) = 952.$$

We have

$$\int_{10}^{26} f(x)\, dx \approx \frac{1240 + 952}{2} = 1096.$$

Solutions for Section 3.3

1. There are 8 whole grid squares and 6 partial grid squares, each of which is about 1/2 a square. The area is about $8(1) + 6\left(\frac{1}{2}\right) = 11.0$

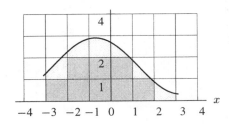

Figure 3.7

5. $\int_0^3 f(x)\,dx$ is equal to the area shaded. We can use Riemann sum to estimate this area, or we can count boxes. These are 3 whole boxes and about 4 half-boxes, for a total of 5 boxes. Since each box represent 4 square units, our estimated area is $5(4) = 20$. We have $\int_0^3 f(x)\,dx \approx 20$. See Figure 3.8.

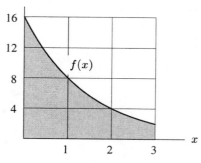

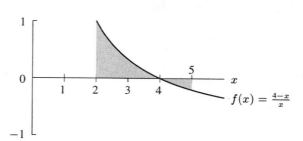

Figure 3.8   **Figure 3.9**

9. (a) Counting the squares yields an estimate of 16.5, each with area $= 1$, so the total shaded area is approximately 16.5.
    (b)

$$\int_0^8 f(x)dx = (\text{shaded area above } x\text{-axis}) - (\text{shaded area below } x\text{-axis})$$

$$\approx 6.5 - 10 = -3.5$$

(c) The answers in (a) and (b) are different because the shaded area below the x-axis is subtracted in order to find the value of the integral in (b).

13. See Figure 3.9. More area appears to be above the x-axis than below it, so the integral is positive.

17. Inspection of the graph in Figure 3.10 tells us that the curves intersect at $(0,0)$ and $(3,9)$, with $3x \geq x^2$ for $0 \leq x \leq 3$, so we can find the area by evaluating the integral

$$\int_0^3 (3x - x^2)dx.$$

Using technology to evaluate the integral, we see

$$\int_0^3 (3x - x^2)dx = 4.5.$$

So the area between the graphs is 4.5.

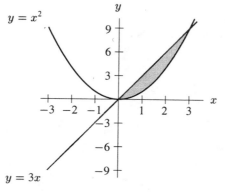

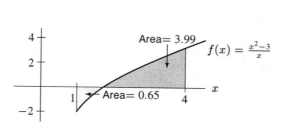

Figure 3.10   **Figure 3.11**

21. Using a calculator or computer, we see that

$$\int_1^4 \frac{x^2 - 3}{x}\, dx \approx 3.34.$$

The graph of $f(x) = \dfrac{x^2 - 3}{x}$ is shown in Figure 3.11. The function is negative to the left of $x = \sqrt{3} \approx 1.73$ and positive to the right of $x = 1.73$. We compute

$$\int_1^{1.73} \frac{x^2 - 3}{x}\, dx = -\text{Area below axis} \approx -0.65$$

and

$$\int_{1.73}^4 \frac{x^2 - 3}{x}\, dx = \text{Area above axis} \approx 3.99.$$

See Figure 3.11. Then

$$\int_1^4 \frac{x^2 - 3}{x}\, dx = \text{Area above axis} - \text{Area below axis} = 3.99 - 0.65 = 3.34.$$

25. (a) 0, since the integrand is an odd function and the limits are symmetric around 0.
     (b) 0, since the integrand is an odd function and the limits are symmetric around 0.

29. (a) We know that $\int_2^5 f(x)\, dx = \int_0^5 f(x)\, dx - \int_0^2 f(x)\, dx$. By symmetry, $\int_0^2 f(x)\, dx = \frac{1}{2}\int_{-2}^2 f(x)\, dx$, so $\int_2^5 f(x)\, dx = \int_0^5 f(x)\, dx - \frac{1}{2}\int_{-2}^2 f(x)\, dx$.
     (b) $\int_2^5 f(x)\, dx = \int_{-2}^5 f(x)\, dx - \int_{-2}^2 f(x)\, dx = \int_{-2}^5 f(x)\, dx - 2\int_{-2}^0 f(x)\, dx$.
     (c) Using symmetry again, $\int_0^2 f(x)\, dx = \frac{1}{2}\left(\int_{-2}^5 f(x)\, dx - \int_2^5 f(x)\, dx\right)$.

Solutions for Section 3.4

1. $\int_a^b f(t)\, dt$ is measured in

$$\left(\frac{\text{miles}}{\text{hours}}\right) \cdot (\text{hours}) = \text{miles}.$$

5. Change in income $= \int_0^{12} r(t)\, dt = \int_0^{12} 40(1.002)^t\, dt = \$485.80$

9. (a) The distance traveled in the first 3 hours (from $t = 0$ to $t = 3$) is given by

$$\int_0^3 (40t - 10t^2)\, dt.$$

     (b) The shaded area in Figure 3.12 represents the distance traveled.

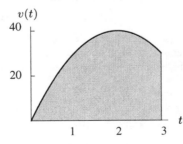

Figure 3.12

     (c) Using a calculator, we get

$$\int_0^3 (40t - 10t^2)\, dt = 90.$$

So the total distance traveled is 90 miles.

13. Notice that the area of a square on the graph represents $\frac{10}{6}$ miles. At $t = 1/3$ hours, $v = 0$. The area between the curve v and the t-axis over the interval $0 \le t \le 1/3$ is $-\int_0^{1/3} v \, dt \approx \frac{5}{3}$. Since v is negative here, she is moving toward the lake. At $t = \frac{1}{3}$, she is about $5 - \frac{5}{3} = \frac{10}{3}$ miles from the lake. Then, as she moves away from the lake, v is positive for $\frac{1}{3} \le t \le 1$. At $t = 1$,

$$\int_0^1 v \, dt = \int_0^{1/3} v \, dt + \int_{1/3}^1 v \, dt \approx -\frac{5}{3} + 8 \cdot \frac{10}{6} = \frac{35}{3},$$

and the cyclist is about $5 + \frac{35}{3} = \frac{50}{3} = 16\frac{2}{3}$ miles from the lake. Since, starting from the moment $t = \frac{1}{3}$, she moves away from the lake, the cyclist will be farthest from the lake at $t = 1$. The maximal distance equals $16\frac{2}{3}$ miles.

17. (a) The amount leaked between $t = 0$ and $t = 2$ is $\int_0^2 R(t) \, dt$.

   (b)

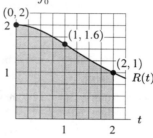

   (c) The rectangular boxes on the diagram each have area $\frac{1}{16}$. Of these 45 are wholly beneath the curve, hence the area under the curve is certainly more than $\frac{45}{16} > 2.81$. There are 9 more partially beneath the curve, and so the desired area is completely covered by 54 boxes. Therefore the area is less than $\frac{54}{16} < 3.38$.

   These are very safe estimates but far apart. We can do much better by estimating what fractions of the broken boxes are beneath the curve. Using this method, we can estimate the area to be about 3.2, which corresponds to 3.2 gallons leaking over two hours.

Solutions for Section 3.5

1. The units for the integral $\int_{800}^{900} C'(q) dq$ are $\left(\frac{\text{dollars}}{\text{tons}}\right) \cdot (\text{tons}) = \text{dollars}$.

   $\int_{800}^{900} C'(q) dq$ represents the cost of increasing production from 800 tons to 900 tons.

5. (a) There are approximately 5.5 squares under the curve of $C'(q)$ from 0 to 30. Each square represents $100, so the total variable cost to produce 30 units is around $550. To find the total cost, we had the fixed cost

$$\text{Total cost} = \text{fixed cost} + \text{total variable cost}$$
$$= 10,000 + 550 = \$10,550.$$

   (b) There are approximately 1.5 squares under the curve of $C'(q)$ from 30 to 40. Each square represents $100, so the additional cost of producing items 31 through 40 is around $150.

   (c) Examination of the graph tells us that $C'(25) = 10$. This means that the cost of producing the 26th item is approximately $10.

9. (a)

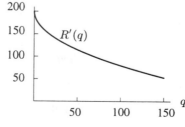

Figure 3.13

(b)   By the Fundamental Theorem,

$$\int_0^{100} R'(q)dq = R(100) - R(0).$$

$R(0) = 0$ because no revenue is produced if no units are sold. Thus we get

$$R(100) = \int_0^{100} R'(q)dq \approx \$12{,}000.$$

(c)   The marginal revenue in selling the 101st unit is given by $R'(100) = \$80/$unit. The total revenue in selling 101 units is:

$$R(100) + R'(100) = \$12{,}080.$$

13.   Since $F'(x)$ is positive for $0 < x < 2$ and $F'(x)$ is negative for $2 < x < 2.5$, $F(x)$ increases on $0 < x < 2$ and decreases on $2 < x < 2.5$. From this we conclude that $F(x)$ has a maximum at $x = 2$. From the process used in Problem 11 we see that the chart agrees with this assumption and that $F(2) = 5.333$.

17.   (a)   Using a calculator, $\int_1^4 3x^2\,dx = 63$.
(b)   Using the Fundamental Theorem of Calculus with $a = 1$ and $b = 4$,

$$\int_1^4 F'(x)dx = F(4) - F(1).$$

Substituting $F(x)$ and $F'(x)$ yields

$$\int_1^4 3x^2 dx = F(4) - F(1) = 4^3 - 1^3 = 64 - 1 = 63$$

which is the exact value of the integral.

Solutions for Chapter 3 Review

1.   The table gives the rate of emissions of nitrogen oxide in millions of metric tons per year. To find the total emissions, we use left-hand and right-hand Riemann sums. We have

$$\text{Left-hand sum} = (6.9)(10) + (9.4)(10) + (13.0)(10) + (18.5)(10) + (20.9)(10) = 687.$$
$$\text{Right-hand sum} = (9.4)(10) + (13.0)(10) + (18.5)(10) + (20.9)(10) + (19.6)(10) = 814.$$
$$\text{Average of left- and right-hand sums} = \frac{687 + 814}{2} = 750.5.$$

The total emissions of nitrogen oxide between 1940 and 1990 is about 750 million metric tons.

5. $\int_0^1 \sqrt{1 + t^2}dt = 1.15.$

9.   Calculating both the LHS and RHS and averaging the two, we get

$$\frac{1}{2}(5(100 + 82 + 69 + 60 + 53) + 5(82 + 69 + 60 + 53 + 49)) = 1692.5$$

13.   (a)

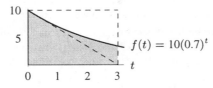

Figure 3.14

Since $f(t)$ is the rate of change of the oil leaking out of the tanker, then the total quantity of oil which has leaked out between $t = 0$ and $t = 3$ is given by $\int_0^3 10(0.7)^t\,dt$, which has the same value as the area under the graph of $f(t)$ between $t = 0$ and $t = 3$. Therefore, the total quantity of oil which leaked out between $t = 0$ and $t = 3$ equals

65% of the area of the rectangle 10 units by 3 units $= (0.65)(3)(10) = 19.5$ gallons.

(b) $\displaystyle\int_0^3 10(0.7)^t \, dt \approx 18.42$ gallons.

(c) Since the area under the graph of $f(t)$ between $t = 0$ and $t = 1$ is greater than the area under the graph of $f(t)$ between $t = 2$ and $t = 3$ we can say that more oil leaked during the first minute than during the third minute.

17. (a)

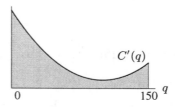

Figure 3.15

The total variable cost of producing 150 units is represented by the area under the graph of $C'(q)$ between 0 and 150, or

$$\int_0^{150} (0.005q^2 - q + 56)dq.$$

(b) An estimate of the total cost of producing 150 units is given by

$$20{,}000 + \int_0^{150} (0.005q^2 - q + 56)dq.$$

This represents the fixed cost ($20,000) plus the variable cost of producing 150 units, which is represented by the integral. Using a calculator, we see

$$\int_0^{150} (0.005q^2 - q + 56)dq \approx 2{,}775.$$

So the total cost is approximately

$$\$20{,}000 + \$2{,}775 = \$22{,}775.$$

(c) $C'(150) = 0.005(150)^2 - 150 + 56 = 18.5$. This means that the marginal cost of the 150th item is 18.5. In other words, the 151st item will cost approximately $18.50.

(d) $C(151)$ is the total cost of producing 151 items. This can be found by adding the total cost of producing 150 items (found in part (b)) and the additional cost of producing the 151st item ($C'(150)$, found in (c)). So we have

$$C(151) \approx 22{,}775 + 18.50 = \$22{,}793.50.$$

21. (a) The acceleration is positive for $0 \le t < 40$ and for a tiny period before $t = 60$, since the slope is positive over these intervals. Just to the left of $t = 40$, it looks like the acceleration is approaching 0. Between $t = 40$ and a moment just before $t = 60$, the acceleration is negative.

(b) The maximum altitude was about 500 feet, when t was a little greater than 40 (here we are estimating the area under the graph for $0 \le t \le 42$).

(c) The total change in altitude for the Montgolfiers and their balloon is the definite integral of their velocity, or the total area under the given graph (counting the part after $t = 42$ as negative, of course). As mentioned before, the total area of the graph for $0 \le t \le 42$ is about 500. The area for $t > 42$ is about 220. So subtracting, we see that the balloon finished 280 feet or so higher than where it began.

25. (a) In the beginning, both birth and death rates are small; this is consistent with a very small population. Both rates begin climbing, the birth rate faster than the death rate, which is consistent with a growing population. The birth rate is then high, but it begins to decrease as the population increases.

(b)

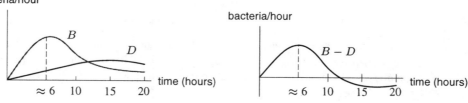

Figure 3.16: Difference between B and D is greatest at $t \approx 6$

The bacteria population is growing most quickly when $B - D$, the rate of change of population, is maximal; that happens when B is farthest above D, which is at a point where the slopes of both graphs are equal. That point is $t \approx 6$ hours.

(c)   Total number born by time t is the area under the B graph from $t = 0$ up to time t. See Figure 3.17.

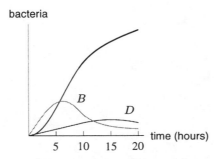

Figure 3.17: Number born by time t is $\int_0^t B(x) \, dx$

Total number alive at time t is the number born minus the number that have died, which is the area under the B graph minus the area under the D graph, up to time t. See Figure 3.18.

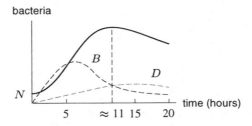

Figure 3.18: Number alive at time t is $\int_0^t (B(x) - D(x)) \, dx$

From Figure 3.18, we see that the population is at a maximum when $B = D$, that is, after about 11 hours. This stands to reason, because $B - D$ is the rate of change of population, so population is maximized when $B - D = 0$, that is, when $B = D$.

CHAPTER FOUR

Solutions for Section 4.1

1. $\dfrac{dy}{dx} = 0$

5. $y' = -12x^{-13}$.

9. $f'(x) = -4x^{-5}$.

13. $y' = 18x^2 + 8x - 2$.

17. $\dfrac{dy}{dq} = 8.4q - 0.5$.

21. $y' = 15t^4 - \frac{5}{2}t^{-1/2} - \frac{7}{t^2}$.

25. $f'(x) = 2x + 3$, so $f'(0) = 3$, $f'(3) = 9$, and $f'(-2) = -1$.

29. $f'(t) = 6t^2 - 8t + 3$ and $f''(t) = 12t - 8$.

33.

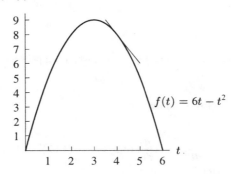

$f(t) = 6t - t^2$

Figure 4.1

To find the equation of a line we need to have a point on the line and its slope. We know that this line is tangent to the curve $f(t) = 6t - t^2$ at $t = 4$. From this we know that both the curve and the line tangent to it will share the same point and the same slope. At $t = 4$, $f(4) = 6(4) - (4)^2 = 24 - 16 = 8$. Thus we have the point $(4, 8)$. To find the slope, we need to find the derivative. The derivative of $f(t)$ is $f'(t) = 6 - 2t$. The slope of the tangent line at $t = 4$ is $f'(4) = 6 - 2(4) = 6 - 8 = -2$. Now that we have a point and the slope, we can find an equation for the tangent line:

$$y = b + mt$$
$$8 = b + (-2)(4)$$
$$b = 16.$$

Thus, $y = -2t + 16$ is the equation for the line tangent to the curve at $t = 4$.

37. (a) The yield is $f(5) = 320 + 140(5) - 10(5)^2 = 770$ bushels per acre.
    (b) $f'(x) = 140 - 20x$, so $f'(5) = 40$ bushels per acre per pound of fertilizer. For each acre, yield will go up by about 40 bushels if an additional pound of fertilizer is used.
    (c) More should be used, because at this level of use, more fertilizer will result in a higher yield. Fertilizer's use should be increased until an additional unit results in a decrease in yield. i.e. until the derivative at that point becomes negative.

41. (a) $f'(x) = 3x^2 - 12 = 3(x - 2)(x + 2)$, which is negative when $-2 < x < 2$. Thus $f(x)$ is decreasing for $-2 < x < 2$.
    (b) $f''(x) = 6x$, which is negative for $x < 0$, so the graph of $f(x)$ is concave down for $x < 0$.
    (c) $f(x)$ is both decreasing and concave down for $-2 < x < 0$.

45. $V = \frac{4}{3}\pi r^3$
     $\frac{dV}{dr} = 4\pi r^2 =$ surface area of a sphere.

   Our reasoning is similar to that of Problem 44. Tne difference quotient $\frac{V(r+h)-V(r)}{h}$ is the volume between two spheres divided by the change in radius. Furthermore, when h is very small (and consequently $V(r+h) \approx V(r)$) this volume is like a coating of paint of depth h applied to the surface of the sphere. The volume of the paint is about $h \cdot$ (Surface Area) for small h: dividing by h gives back the surface area. Also, thinking about the derivative as the rate of change of the function for a small change in the variable, the answer seems clear. If you increase the radius of a sphere the tiniest amount, the volume will increase by a very thin layer whose volume will be the surface area at that radius multiplied by that tiniest amount.

Solutions for Section 4.2

1. $y' = 10t + 4e^t$.

5. $\frac{dy}{dx} = 3 - 2(\ln 4)4^x$.

9. $P'(t) = Ce^t$.

13. $y' = Ae^t$

17. $P'(t) = 12.41(\ln 0.94)(0.94)^t$.

21. $Ae^t + \frac{B}{t}$.

25. (a) $f(x) = 1 - e^x$ crosses the x-axis where $0 = 1 - e^x$, which happens when $e^x = 1$, so $x = 0$. Since $f'(x) = -e^x$, $f'(0) = -e^0 = -1$.
     (b) $y = -x$

29. (a) $P = 4.1(1 + 0.02)^t = 4.1(1.02)^t$ billion.
     (b)

$$\frac{dP}{dt} = 4.1\frac{d}{dt}(1.02)^t = 4.1(1.02)^t(\ln 1.02).$$

$$\left.\frac{dP}{dt}\right|_{t=0} = 4.1(1.02)^0 \ln 1.02 \approx 0.0812 \text{ billion people per year.}$$

$$\left.\frac{dP}{dt}\right|_{t=15} = 4.1(1.02)^{15} \ln 1.02 \approx 0.1093 \text{ billion people per year.}$$

$\frac{dP}{dt}$ is the rate of growth of the world's population; $\left.\frac{dP}{dt}\right|_{t=0}$ and $\left.\frac{dP}{dt}\right|_{t=15}$ are the rates of growth in the years 1975 and 1990, respectively.

33. We are interested in when the derivative $\frac{d(a^x)}{dx}$ is positive and when it is negative. The quantity a^x is always positive. However $\ln a > 0$ for $a > 1$ and $\ln a < 0$ for $0 < a < 1$. Thus the function a^x is increasing for $a > 1$ and decreasing for $a < 1$.

37. (a) $V(4) = 25(0.85)^4 = 25(0.522) = 13,050$. Thus the value of the car after 4 years is $13,050.
     (b) We have a function of the form $f(t) = Ca^t$. We know that such functions have a derivative of the form $(C \ln a) \cdot a^t$. Thus, $V'(t) = 25(0.85)^t \cdot \ln 0.85 = -4.063(0.85)^t$. The units would be the change in value (in thousands of dollars) with respect to time (in years), or thousands of dollars/year.
     (c) $V'(4) = -4.063(0.85)^4 = -4.063(0.522) = -2.121$. This means that at the end of the fourth year, the value of the car is decreasing by $2121 per year.
     (d) $V(t)$ is a positive decreasing function, so that the value of the automobile is positive and decreasing. $V'(t)$ is a negative function whose magnitude is decreasing, meaning the value of the automobile is always dropping, but the yearly loss of value is less as time goes on. The graphs of $V(t)$ and $V'(t)$ confirm that the value of the car decreases with time. What they do not take into account are the *costs* associated with owning the vehicle. At some time, t, it is likely that the costs of owning the vehicle will outweigh its value. At that time, it may no longer be worthwhile to keep the car.

Solutions for Section 4.3

1. $f'(x) = 99(x + 1)^{98} \cdot 1 = 99(x + 1)^{98}$.

5. We need to find $\dfrac{dw}{dr}$. Let's begin by setting $u = 5r - 6$. We now have $w = u^3$. We know that $\dfrac{dw}{dr} = \dfrac{dw}{du} \cdot \dfrac{du}{dr}$. Now $\dfrac{dw}{du} = 3u^2$, while $\dfrac{du}{dr} = 5$. Thus, substituting $5r - 6$ back in for u, we get

$$\frac{dw}{dr} = \frac{dw}{du} \cdot \frac{du}{dr} = 3(5r - 6)^2 \cdot 5 = 15(5r - 6)^2.$$

9. $y' = \dfrac{3s^2}{2\sqrt{s^3 + 1}}$.

13. $\frac{dy}{dt} = \frac{5}{5t+1}$.

17. By the chain rule,

$$\frac{dC}{dq} = (12)((3)(3q^2 - 5)^2)(6q) = 216q(3q^2 - 5)^2.$$

21. $f'(t) = \frac{2t}{t^2+1}$.

25. $g'(t) = \dfrac{1}{4t + 9}(4) = \dfrac{4}{4t + 9}$.

29. $\dfrac{dy}{dx} = 2(5 + e^x)e^x$.

33. $f(p) = 10,000e^{-0.25p}$, $f(2) = 10,000e^{-0.5} \approx 6065$. If the product sells for $2, then 6065 units can be sold.

$$f'(p) = 10,000e^{-0.25p}(-0.25) = -2500e^{-0.25p}$$
$$f'(2) = -2,500e^{-0.5} \approx -1516$$

$f'(2) = -1516$ means that at a price of $2, a $1 increase in price will result in a decrease in quantity sold of 1516 units .

37. (a)
$$\frac{dH}{dt} = \frac{d}{dt}(40 + 30e^{-2t}) = 30(-2)e^{-2t} = -60e^{-2t}.$$

(b) Since e^{-2t} is always positive, $\dfrac{dH}{dt} < 0$; this makes sense because the temperature of the soda is decreasing.

(c) The magnitude of $\dfrac{dH}{dt}$ is

$$\left|\frac{dH}{dt}\right| = \left|-60e^{-2t}\right| = 60e^{-2t} \le 60 = \left|\frac{dH}{dt}_{t=0}\right|,$$

since $e^{-2t} \le 1$ for all $t \ge 0$ and $e^0 = 1$. This is just saying that at the moment that the can of soda is put in the refrigerator (at $t = 0$), the temperature difference between the soda and the inside of the refrigerator is the greatest, so the temperature of the soda is dropping the quickest.

Solutions for Section 4.4

1. By the product rule, $f'(x) = 2x(x^3 + 5) + x^2(3x^2) = 2x^4 + 3x^4 + 10x = 5x^4 + 10x$. Alternatively, $f'(x) = (x^5 + 5x^2)' = 5x^4 + 10x$. The two answers should, and do, match.

5. $y' = 2^x + x(\ln 2)2^x = 2^x(1 + x \ln 2)$.

9. $w' = (3t^2 + 5)(t^2 - 7t + 2) + (t^3 + 5t)(2t - 7)$.

13. $\dfrac{dP}{dt} = (t^2)(\frac{1}{t}) + (2t)(\ln t) = t + 2t \ln t$.

17. $f'(z) = \dfrac{1}{2\sqrt{z}}e^{-z} - \sqrt{z}e^{-z}.$

21. $f'(x) = \dfrac{e^x \cdot 1 - x \cdot e^x}{(e^x)^2} = \dfrac{e^x(1-x)}{(e^x)^2} = \dfrac{1-x}{e^x}.$

25.   Using the quotient rule,

$$\frac{dw}{dy} = \frac{d}{dy}\left(\frac{3y + y^2}{5+y}\right) = \frac{(3+2y)(5+y) - (3y+y^2)\cdot 1}{(5+y)^2}$$

$$= \frac{15 + 13y + 2y^2 - 3y - y^2}{(5+y)^2} = \frac{15 + 10y + y^2}{(5+y)^2}.$$

29.   (a)

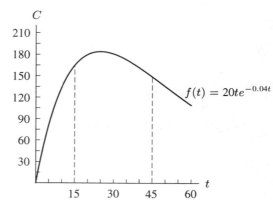

Figure 4.2

Looking at the graph of C, we can see that the see that at $t = 15$, C is increasing. Thus, the slope of the curve at that point is positive, and so $f'(15)$ is also positive. At $t = 45$, the function is decreasing, i.e. the slope of the curve is negative, and thus $f'(45)$ is negative.

(b)   We begin by differentiating the function:

$$f'(t) = (20t)(-0.04e^{-0.04t}) + (e^{-0.04t})(20)$$
$$f'(t) = e^{-0.04t}(20 - 0.8t).$$

At $t = 30$,

$$f(30) = 20(30)e^{-0.04 \cdot (30)} = 600e^{-1.2} \approx 181 \text{ mg/ml}$$
$$f'(30) = e^{-1.2}(20 - (0.8)(30)) = e^{-1.2}(-4) \approx -1.2 \text{ mg/ml/min}.$$

These results mean the following: At $t = 30$, or after 30 minutes, the concentration of the drug in the body ($f(30)$) is about 181 mg/ml. The rate of change of the concentration ($f'(30)$) is about -1.2 mg/ml/min, meaning that the concentration of the drug in the body is dropping by 1.2 mg/ml each minute at $t = 30$ minutes.

Solutions for Section 4.5

1. $\dfrac{dy}{dx} = 5\cos x.$

5. $\dfrac{dy}{dx} = 5\cos x - 5.$

9. $\dfrac{dy}{dt} = A(\cos(Bt)) \cdot B = AB\cos(Bt).$

13. $f'(x) = \cos(3x) \cdot 3 = 3\cos(3x).$

17.   Using the quotient rule, we get

$$\frac{d}{dt}\left(\frac{t^2}{\cos t}\right) = \frac{2t\cos t - t^2(-\sin t)}{(\cos t)^2}$$

$$= \frac{2t\cos t + t^2\sin t}{(\cos t)^2}.$$

21.   (a) $v(t) = \dfrac{dy}{dt} = \dfrac{d}{dt}(15 + \sin(2\pi t)) = 2\pi\cos(2\pi t).$

(b)

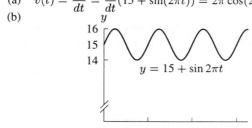

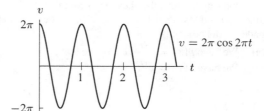

Solutions for Chapter 4 Review

1. $f'(t) = 24t^3.$

5. $f'(x) = 3x^2 - 6x + 5.$

9. $f'(x) = 2x + \frac{3}{x}.$

13. $f'(z) = \frac{1}{z^2+1}(2z) = \frac{2z}{z^2+1}.$

17. $s'(t) = 2t + \frac{2}{t}.$

21. $f'(x) = 2\cos(2x).$

25. $\dfrac{dz}{dt} = \dfrac{3(5t+2) - (3t+1)5}{(5t+2)^2} = \dfrac{15t+6-15t-5}{(5t+2)^2} = \dfrac{1}{(5t+2)^2}.$

29.   (a) $H'(2) = r'(2) + s'(2) = -1 + 3 = 2.$

(b) $H'(2) = 5s'(2) = 5(3) = 15.$

(c) $H'(2) = r'(2)s(2) + r(2)s'(2) = -1 \cdot 1 + 4 \cdot 3 = 11.$

(d) $H'(2) = \dfrac{r'(2)}{2\sqrt{r(2)}} = \dfrac{-1}{2\sqrt{4}} = -\dfrac{1}{4}.$

33.   If $a = e$, the only solution is $(0,1)$.

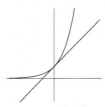

   If $1 < a < e$, there are two solutions as illustrated:

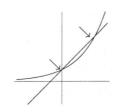

and if $a > e$, there are also two solutions.

One way to prove the above is to compare the slopes of the lines. For example, e^x will have slope greater than 1 for all $x > 0$ and less than 1 for all $x < 0$, so it cannot meet the line $1 + x$ at any other points. Similar arguments can be made for the other cases.

37. (a) We have $p(x) = x^2 - x$. We see that $p'(x) = 2x - 1 < 0$ when $x < \frac{1}{2}$. So p is decreasing when $x < \frac{1}{2}$.

  (b) We have $p(x) = x^{1/2} - x$, so

$$p'(x) = \frac{1}{2}x^{-1/2} - 1 < 0$$
$$\frac{1}{2}x^{-1/2} < 1$$
$$x^{-1/2} < 2$$
$$x^{1/2} > \frac{1}{2}$$
$$x > \frac{1}{4}.$$

Thus $p(x)$ is decreasing when $x > \frac{1}{4}$.

  (c) We have $p(x) = x^{-1} - x$, so

$$p'(x) = -1x^{-2} - 1 < 0$$
$$-x^{-2} < 1$$
$$x^{-2} > -1,$$

which is always true where x^{-2} is defined since $x^{-2} = 1/x^2$ is always positive. Thus $p(x)$ is decreasing for $x < 0$ and for $x > 0$.

41. All of the functions go through the origin. They will look the same if they have the same tangent line, or equivalently, the same slope at $x = 0$. Therefore for each function we find the derivative and evaluate it at $x = 0$:

$$
\begin{array}{lll}
\text{For } y = x, & y' = 1, & \text{so } y'(0) = 1. \\
\text{For } y = \sqrt{x}, & y' = \frac{1}{2\sqrt{x}}, & \text{so } y'(0) \text{ is undefined.} \\
\text{For } y = x^2, & y' = 2x, & \text{so } y'(0) = 0. \\
\text{For } y = x^3 + \frac{1}{2}x^2, & y' = 3x^2 + x, & \text{so } y'(0) = 0. \\
\text{For } y = x^3, & y' = 3x^2, & \text{so } y'(0) = 0. \\
\text{For } y = \ln(x + 1), & y' = \frac{1}{x+1}, & \text{so } y'(0) = 1. \\
\text{For } y = \frac{1}{2}\ln(x^2 + 1), & y' = \frac{x}{x^2+1}, & \text{so } y'(0) = 0. \\
\text{For } y = \sqrt{2x - x^2}, & y' = \frac{1-x}{\sqrt{2x-x^2}}, & \text{so } y'(0) \text{ is undefined.}
\end{array}
$$

So near the origin, functions with $y'(0) = 1$ will all be indistinguishable resembling the line $y = x$. These functions are:

$$y = x \quad \text{and} \quad y = \ln(x + 1).$$

Functions with $y'(0) = 0$ will be indistinguishable near the origin and resemble the line $y = 0$ (a horizontal line). These functions are:

$$y = x^2, \qquad y = x^3 + \frac{1}{2}x^2, \qquad y = x^3, \qquad \text{and} \qquad y = \frac{1}{2}\ln(x^2 + 1).$$

Functions that have undefined derivatives at $x = 0$ look like vertical lines at the origin. These functions are

$$y = \sqrt{x} \quad \text{and} \quad y = \sqrt{2x - x^2}.$$

45. (a) $P(12) = 10e^{0.6(12)} = 10e^{7.2} \approx 13{,}394$ zebra mussels. There are 13,394 zebra mussels in the area after 12 months.

(b) We differentiate to find $P'(t)$, and then substitute in to find $P'(12)$:

$$P'(t) = 10(e^{0.6t})(0.6) = 6e^{0.6t}$$

$$P'(12) = 6e^{0.6(12)} \approx 8{,}037 \text{ mussels/month.}$$

The population is growing at a rate of approximately 8037 zebra mussels per month.

Solutions to Problems on Establishing the Derivative Formulas

1. Using the definition of the derivative, we have

$$\begin{aligned} f'(x) &= \lim_{h \to 0} \frac{f(x+h) - f(x)}{h} \\ &= \lim_{h \to 0} \frac{2(x+h) + 1 - (2x+1)}{h} \\ &= \lim_{h \to 0} \frac{2x + 2h + 1 - 2x - 1}{h} \\ &= \lim_{h \to 0} \frac{2h}{h}. \end{aligned}$$

As long as h is very close to, but not actually equal to, zero we can say that $\lim_{h \to 0} \dfrac{2h}{h} = 2$, and thus conclude that $f'(x) = 2$.

5. The definition of the derivative states that

$$f'(x) = \lim_{h \to 0} \frac{f(x+h) - f(x)}{h}.$$

Using this definition, we have

$$\begin{aligned} f'(x) &= \lim_{h \to 0} \frac{4(x+h)^2 + 1 - (4x^2 + 1)}{h} \\ &= \lim_{h \to 0} \frac{4x^2 + 8xh + 4h^2 + 1 - 4x^2 - 1}{h} \\ &= \lim_{h \to 0} \frac{8xh + 4h^2}{h} \\ &= \lim_{h \to 0} \frac{h(8x + 4h)}{h}. \end{aligned}$$

As long as h approaches, but does not equal, zero we can cancel it out of the numerator and denominator. The derivative now becomes

$$\lim_{h \to 0} (8x + 4h) = 8x.$$

Thus, $f'(x) = 6x$ as we stated above.

Solutions to Practice Problems on Differentiation

1. $f'(t) = 2t + 4t^3$

5. $f'(x) = -2x^{-1} + 5\left(\frac{1}{2}x^{-1/2}\right) = \dfrac{-2}{x} + \dfrac{5}{2\sqrt{x}}$

9. $D'(p) = 2pe^{p^2} + 10p$

13. $s'(t) = \dfrac{16}{2t+1}$

17. $C'(q) = 3(2q+1)^2 \cdot 2 = 6(2q+1)^2$

21. $y' = 2x \ln(2x+1) + \dfrac{2x^2}{2x+1}$

25. $g'(t) = 15 \cos (5t)$

29. $y' = 17 + 12x^{-1/2}$.

33.   Either notice that $f(x) = \dfrac{x^2 + 3x + 2}{x + 1}$ can be written as $f(x) = \dfrac{(x + 2)(x + 1)}{x + 1}$ which reduces to $f(x) = x + 2$, giving $f'(x) = 1$, or use the quotient rule which gives

$$\begin{aligned} f'(x) &= \frac{(x + 1)(2x + 3) - (x^2 + 3x + 2)}{(x + 1)^2} \\ &= \frac{2x^2 + 5x + 3 - x^2 - 3x - 2}{(x + 1)^2} \\ &= \frac{x^2 + 2x + 1}{(x + 1)^2} \\ &= \frac{(x + 1)^2}{(x + 1)^2} \\ &= 1. \end{aligned}$$

37. $q'(r) = \dfrac{3(5r + 2) - 3r(5)}{(5r + 2)^2} = \dfrac{15r + 6 - 15r}{(5r + 2)^2} = \dfrac{6}{(5r + 2)^2}$

41. $h'(w) = 5(w^4 - 2w)^4(4w^3 - 2)$

45. $h'(w) = 6w^{-4} + \dfrac{3}{2}w^{-1/2}$

49.   Using the chain rule, $g'(\theta) = (\cos \theta)e^{\sin \theta}$.

53. $h'(r) = \dfrac{d}{dr}\left(\dfrac{r^2}{2r + 1}\right) = \dfrac{(2r)(2r + 1) - 2r^2}{(2r + 1)^2} = \dfrac{2r(r + 1)}{(2r + 1)^2}$.

57. $f'(x) = \dfrac{3x^2}{9}(3 \ln x - 1) + \dfrac{x^3}{9}\left(\dfrac{3}{x}\right) = x^2 \ln x - \dfrac{x^2}{3} + \dfrac{x^2}{3} = x^2 \ln x$

61.   Using the quotient rule gives

$$\begin{aligned} w'(r) &= \frac{2ar(b + r^3) - 3r^2(ar^2)}{(b + r^3)^2} \\ &= \frac{2abr - ar^4}{(b + r^3)^2}. \end{aligned}$$

CHAPTER FIVE

Solutions for Section 5.1

1. We find a critical point by noting where $f'(x) = 0$ or f' is undefined. Since the curve is smooth throughout, f' is always defined, so we look for where $f'(x) = 0$, or equivalently where the tangent line to the graph is horizontal. These points are shown in Figure 5.1:

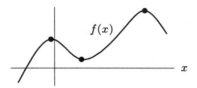

Figure 5.1

As we can see, there are three critical points. The leftmost one is a local maximum, because points near it are all lower; similarly, the middle critical point is surrounded by higher points, and is a local minimum. The critical point to the right is a local maximum.

5. (a)   (b)

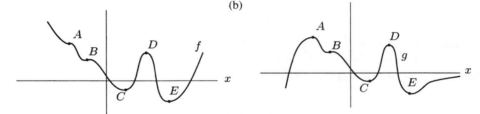

9. A critical point of f requires $f'(x) = 0$ or f' undefined. Since f' is clearly defined over the relevant range, we find where $f'(x) = 0$, that is, where the graph of f' crosses the x-axis. These points are shown and labeled in Figure 5.2:

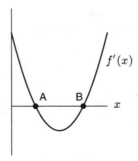

Figure 5.2

To the left of critical point A, we see that $f' > 0$ and f is increasing; to the right of the critical point, we see that $f' < 0$ and f is decreasing. So there is a local maximum at A.

To the left of critical point B, we see that $f' < 0$ and f is decreasing; to the right of the critical point, we see that $f' > 0$ and f is increasing. So there is a local minimum at B. The sketch of $f(x)$ in Figure 5.3 shows A is a local maximum and B is a local minimum:

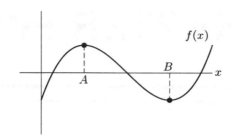

Figure 5.3

13.   If the minimum of $f(x)$ is at $(-2, -3)$, then the derivative of f must be equal to 0 there. In other words, $f'(-2) = 0$. If

$$f(x) = x^2 + ax + b, \quad \text{then}$$
$$f'(x) = 2x + a$$
$$f'(-2) = 2(-2) + a = -4 + a = 0$$

so $a = 4$. Since $(-2, -3)$ is on the graph of $f(x)$ we know that $f(-2) = -3$. So

$$f(-2) = (-2)^2 + a(-2) + b = -3$$
$$a = 4, \text{ so} \qquad (-2)^2 + 4(-2) + b = -3$$
$$4 - 8 + b = -3$$
$$-4 + b = -3$$
$$b = 1$$

so $a = 4$ and $b = 1$, and $f(x) = x^2 + 4x + 1$.

17.   Local maximum for some θ, with $1.1 < \theta < 1.2$
      Local minimum for some θ, with $1.5 < \theta < 1.6$
      Local maximum for some θ, with $2.0 < \theta < 2.1$

Solutions for Section 5.2

1.   We find an inflection point by noting where the concavity changes, or equivalently where the tangent line passes from above the graph to below or vice versa. Such points are shown below in Figure 5.4:

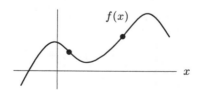

Figure 5.4

There are two inflection points.

5.   From the graph of $f(x)$ in the figure below, we see that the function must have two inflection points. We calculate $f'(x) = 4x^3 + 3x^2 - 6x$, and $f''(x) = 12x^2 + 6x - 6$. Solving $f''(x) = 0$ we find that:

$$x_1 = -1 \quad \text{and} \quad x_2 = \frac{1}{2}.$$

Since $f''(x) > 0$ for $x < x_1$, $f''(x) < 0$ for $x_1 < x < x_2$, and $f''(x) > 0$ for $x_2 < x$, it follows that both points are inflection points. See Figure 5.5.

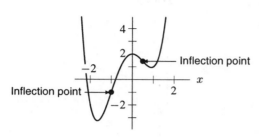

Figure 5.5

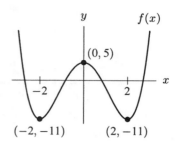

Figure 5.6

9. The derivative of $f(x)$ is $f'(x) = 4x^3 - 16x$. The critical points of $f(x)$ will be points for which $f'(x) = 0$. Factoring, we get

$$4x^3 - 16x = 0$$
$$4x(x^2 - 4) = 0$$
$$4x(x - 2)(x + 2) = 0$$

So the critical points of f will be $x = 0, x = 2,$ and $x = -2$. We see that

$$\begin{aligned} f'(x) &< 0 && \text{for } x < -2 \\ f'(x) &> 0 && \text{for } -2 < x < 0 \\ f'(x) &< 0 && \text{for } 0 < x < 2 \\ f'(x) &> 0 && \text{for } x > 2 \end{aligned}$$

So we conclude that $f(x)$ has local minima at $x = -2$ and $x = 2$, and has a local maximum at $x = 0$. From the graph in Figure 5.6, we see that this is correct.

To find the inflection points of f we look for the points at which $f''(x)$ changes sign. At any such point $f''(x)$ is either zero or undefined. Since $f''(x) = 12x^2 - 16$ our candidate points are $x = \pm 2/\sqrt{3}$. At both of these points $f''(x)$ changes sign, so both of these points are inflection points.

13. (a) One possible answer is shown in Figure 5.7.

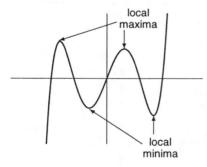

Figure 5.7

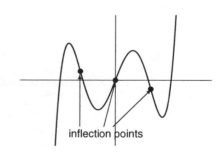

Figure 5.8

(b) This function is concave down at each local maximum and concave up at each local minimum, so it changes concavity at least three times. This function has at least 3 inflection points. See Figure 5.8

17.   We have $f'(x) = 3x^2 - 36x - 10$ and $f''(x) = 6x - 36$. The inflection point occurs where $f''(x) = 0$, hence $6x - 36 = 0$. The inflection point is at $x = 6$. A graph is shown in Figure 5.9.

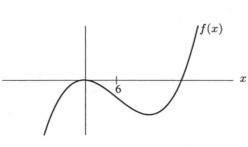

Figure 5.9

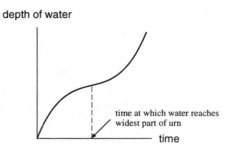

Figure 5.10

21.   See Figure 5.10.

25.   The local maxima and minima of f correspond to places where f' is zero and changes sign or, possibly, to the endpoints of intervals in the domain of f. The points at which f changes concavity correspond to local maxima and minima of f'. The change of sign of f', from positive to negative corresponds to a maximum of f and change of sign of f' from negative to positive corresponds to a minimum of f.

29.

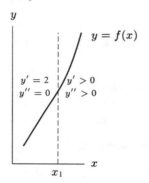

Solutions for Section 5.3

1.   See Figure 5.11.

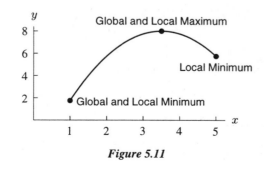

Figure 5.11

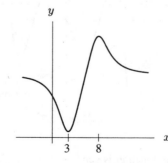

Figure 5.12

5.   See Figure 5.12.

9.

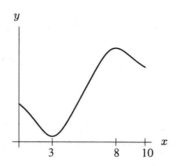

Figure 5.13

13. (a) $f'(x) = 1 - 1/x$. This is zero only when $x = 1$. Now $f'(x)$ is positive when $1 < x \le 2$, and negative when $0.1 < x < 1$. Thus $f(1) = 1$ is a local minimum. The endpoints $f(0.1) \approx 2.4026$ and $f(2) \approx 1.3069$ are local maxima.
    (b) Comparing values of f shows that $x = 0.1$ gives the global maximum and $x = 1$ gives the global minimum.

17. First find the marginal revenue and marginal cost. Note that each product sells for \$588, so revenue is given by $R(q) = 588q$.

$$MR = R'(q) = 588$$
$$MC = C'(q) = 3q^2 - 120q + 1200$$

Setting $MR = MC$ yields

$$3q^2 - 120q + 1200 = 588$$
$$\text{so} \quad 3q^2 - 120q + 612 = 0$$

This factors to

$$3(q - 34)(q - 6) = 0$$

so $MR = MC$ at $q = 34$ and $q = 6$. We now find the profit at these points:

$$R(6) - C(6) = 588(6) - \left[(6)^3 - 60(6)^2 + 1200(6) + 1000\right]$$
$$= 3{,}528 - 6{,}256 = -\$2{,}728$$

$$R(34) - C(34) = 588(34) - \left[(34)^3 - 60(34)^2 + 1200(34) + 1000\right]$$
$$= 19{,}992 - 11{,}744 = \$8{,}248$$

We must also try the endpoints

$$R(0) - C(0) = 588(0) - \left[(0)^3 - 60(0)^2 + 1200(0) + 1000\right]$$
$$= -\$1{,}000$$

$$R(50) - C(50) = 588(50) - \left[(50)^3 - 60(50)^2 + 1200(50) + 1000\right]$$
$$= 29{,}400 - 36{,}000 = -\$6{,}600$$

From this we see that profit is maximized at $q = 34$ units. The total cost at $q = 34$ is $C(34) = \$11{,}744$. The total revenue at $q = 34$ is $R(34) = \$19{,}992$ and the total profit is

$$19{,}992 - 11{,}744 = \$8{,}248.$$

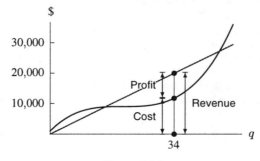

Figure 5.14

21. (a) The fixed cost is 0 because $C(0) = 0$.

(b) Profit, $\pi(q)$, is equal to money from sales, $7q$, minus total cost to produce those items, $C(q)$.

$$\pi = 7q - 0.01q^3 + 0.6q^2 - 13q$$
$$\pi' = -0.03q^2 + 1.2q - 6$$

$$\pi' = 0 \quad \text{if} \quad q = \frac{-1.2 \pm \sqrt{(1.2)^2 - 4(0.03)(6)}}{-0.06} \approx 5.9 \quad \text{or} \quad 34.1.$$

Now $\pi'' = -0.06q + 1.2$, so $\pi''(5.9) > 0$ and $\pi''(34.1) < 0$. This means $q = 5.9$ is a local min and $q = 34.1$ a local max. We now evaluate the endpoint, $\pi(0) = 0$, and the points nearest $q = 34.1$ with integer q-values:

$$\pi(35) = 7(35) - 0.01(35)^3 + 0.6(35)^2 - 13(35) = 245 - 148.75 = 96.25,$$

$$\pi(34) = 7(34) - 0.01(34)^3 + 0.6(34)^2 - 13(34) = 238 - 141.44 = 96.56.$$

So the (global) maximum profit is $\pi(34) = 96.56$. The money from sales is \$238, the cost to produce the items is \$141.44, resulting in a profit of \$96.56.

(c) The money from sales is equal to price$\times$quantity sold. If the price is raised from \$7 by \$$x$ to \$$(7+x)$, the result is a reduction in sales from 34 items to $(34 - 2x)$ items. So the result of raising the price by \$$x$ is to change the money from sales from $(7)(34)$ to $(7 + x)(34 - 2x)$ dollars. If the production level is fixed at 34, then the production costs are fixed at \$141.44, as found in part (b), and the profit is given by:

$$\pi(x) = (7 + x)(34 - 2x) - 141.44$$

This expression gives the profit as a function of change in price x, rather than as a function of quantity as in part (b). We set the derivative of π with respect to x equal to zero to find the change in price that maximizes the profit:

$$\frac{d\pi}{dx} = (1)(34 - 2x) + (7 + x)(-2) = 20 - 4x = 0$$

So $x = 5$, and this must give a maximum for $\pi(x)$ since the graph of π is a parabola which opens downwards. The profit when the price is \$12 $(= 7 + x = 7 + 5)$ is thus $\pi(5) = (7 + 5)(34 - 2(5)) - 141.44 = \$146.56$. This is indeed higher than the profit when the price is \$7, so the smart thing to do is to raise the price by \$5.

25. Let x equal the number of chairs ordered in excess of 300, so $0 \leq x \leq 100$.

$$\text{Revenue} = R = (90 - 0.25x)(300 + x)$$
$$= 27{,}000 - 75x + 90x - 0.25x^2 = 27{,}000 + 15x - 0.25x^2$$

At a critical point $dR/dx = 0$. Since $dR/dx = 15 - 0.5x$, we have $x = 30$, and the maximum revenue is \$27,225 since the graph of R is a parabola which opens downwards. The minimum is \$0 (when no chairs are sold).

29. (a) We know that Profit = Revenue − Cost, so differentiating with respect to q gives:

$$\text{Marginal Profit} = \text{Marginal Revenue} - \text{Marginal Cost}.$$

We see from the figure in the problem that just to the left of $q = a$, marginal revenue is less than marginal cost, so marginal profit is negative there. To the right of $q = a$ marginal revenue is greater than marginal cost, so marginal profit is positive there. At $q = a$ marginal profit changes from negative to positive. This means that profit is decreasing to the left of a and increasing to the right. The point $q = a$ corresponds to a local minimum of profit, and does not maximize profit. It would be a terrible idea for the company to set its production level at $q = a$.

(b) We see from the figure in the problem that just to the left of $q = b$ marginal revenue is greater than marginal cost, so marginal profit is positive there. Just to the right of $q = b$ marginal revenue is less than marginal cost, so marginal profit is negative there. At $q = b$ marginal profit changes from positive to negative. This means that profit is increasing to the left of b and decreasing to the right. The point $q = b$ corresponds to a local maximum of profit. In fact, since the area between the MC and MR curves in the figure in the text between $q = a$ and $q = b$ is bigger than the area between $q = 0$ and $q = a$, $q = b$ is in fact a global maximum.

Solutions for Section 5.4

1. (a) (i)   The average cost of quantity q is given by the formula $a(q)/q$. So average cost at $q = 30$ is given by $C(30)/30$. From the graph, we see that $C(30) \approx 260$, so a(q) $\approx \frac{260}{30} \approx \$8.67$ per unit. To interpret this graphically, note that $a(q) = \frac{C(q)}{q} = \frac{C(q)-0}{q-0}$. This is exactly the formula for the slope of a line from the origin to a point $(q, C(q))$ on the curve. So a(30) is the slope of a line connecting $(0, 0)$ to $(30, C(30))$. Such a line is shown below in Figure 5.15.

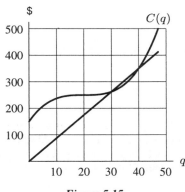

Figure 5.15

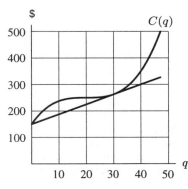

Figure 5.16

(ii)   The marginal cost is given by $C'(q)$. This derivative is equal to the slope of a tangent line to $C(q)$ at $q = 30$. To estimate this slope, it will be easiest to draw the tangent line, as shown in Figure 5.16. From this plot, we see approximately that the point $(40, 300)$ and the point $(0, 160)$ are on this line, so its slope is approximately $\frac{300-160}{40-0} = 3.5$. $C'(30) \approx \$3.50$ per unit.

(b)   We know that $a(q)$ is minimized where $a(q) = C'(q)$. Using the graphical interpretations froms parts (b) and (c), this is equivalent to saying that the tangent line has the same slope as the line connecting the point on the curve to the origin. Since these two lines share a point, specifically the point $(q, C(q))$ on the curve, and have the same slope, they are in fact the same line. So $a(q)$ is minimized where the line passing from $(q, C(q))$ to the origin is also tangent to the curve. To find such points, a variety of lines passing through the origin and the curve are shown in Figure 5.17:

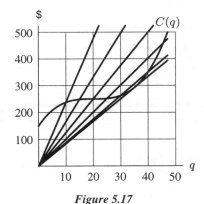

Figure 5.17

From this plot, we see that the line with the desired properties intersects the curve at $q \approx 35$. So $q \approx 35$ units minimizes $a(q)$.

5. (a)   The marginal cost tells us that additional units produced would cost about $10 each, which is below the average cost, so producing them would reduce average cost.

(b)   It is impossible to determine the effect on profit from the information given. Profit depends on both cost and revenue, $\pi = R - C$, but we have no information on revenue.

9. The graph of the average cost function is shown in Figure 5.18.

Average Cost

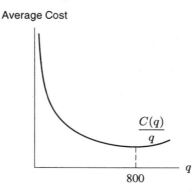

$\dfrac{C(q)}{q}$

800

q

Figure 5.18

13. (a) Since the graph is concave down, the average cost gets smaller as q increases. This is because the cost per item gets smaller as q increases. There is no value of q for which the average cost is minimized since for any q_0 larger than q the average cost at q_0 is less than the average cost at q. Graphically, the average cost at q is the slope of the line going through the origin and through the point $(q, C(q))$. Figure 5.19 shows how as q gets larger, the average cost decreases.

C (cost)

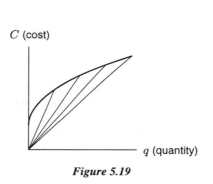

q (quantity)

Figure 5.19

C (cost)

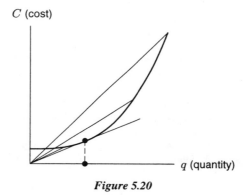

q (quantity)

Figure 5.20

(b) The average cost will be minimized at some q for which the line through $(0, 0)$ and $(q, c(q))$ is tangent to the cost curve. This point is shown in Figure 5.20.

Solutions for Section 5.5

1. The effect on demand is approximately E times the change in price. A price increase cause a decrease in demand and a price decrease cause an increase in demand.

    (a) Demand decreases by about $2(3\%) = 6\%$.
    (b) Demand increases by about $2(3\%) = 6\%$.

5. Table 5.5 of Section 5.5 gives the elasticity of milk as 0.31. Here $0 \leq E \leq 1$, so the demand is inelastic, meaning that changes in price will not change the demand so much. This is expected for milk; as a staple food item, people will continue to buy it even if the price goes up.

9. (a) There were good substitutes for slaves in city occupations - including free blacks. There were no good substitutes in the countryside.
    (b) They were from the countryside, where there was no satisfactory substitute for slaves.

13. (a) We use

$$E = \left| \frac{p}{q} \cdot \frac{dq}{dp} \right|.$$

So we approximate dq/dp at $p = 1.00$.

$$\frac{dq}{dp} \approx \frac{2440 - 2765}{1.25 - 1.00} = \frac{-325}{0.25} = -1300$$

so $$E = \left| \frac{p}{q} \cdot \frac{dq}{dp} \right| \approx \left| \frac{1.00}{2765} \cdot (-1300) \right| = 0.470$$

Since $E = 0.470 < 1$, demand for the candy is inelastic at $p = 1.00$.

(b) At $p = 1.25$,

$$\frac{dq}{dp} \approx \frac{1980 - 2440}{1.50 - 1.25} = \frac{-460}{0.25} = -1840$$

so $$E = \left| \frac{p}{q} \cdot \frac{dq}{dp} \right| \approx \left| \frac{1.25}{2440} \cdot (-1840) \right| = 0.943$$

At $p = 1.5$,

$$\frac{dq}{dp} \approx \frac{1660 - 1980}{1.75 - 1.50} = \frac{-320}{0.25} = -1280$$

so $$E = \left| \frac{p}{q} \cdot \frac{dq}{dp} \right| \approx \left| \frac{1.50}{1980} \cdot (-1280) \right| = 0.970$$

At $p = 1.75$,

$$\frac{dq}{dp} \approx \frac{1175 - 1660}{2.00 - 1.75} = \frac{-485}{0.25} = -1940$$

so $$E = \left| \frac{p}{q} \cdot \frac{dq}{dp} \right| \approx \left| \frac{1.75}{1660} \cdot (-1940) \right| = 2.05$$

At $p = 2.00$,

$$\frac{dq}{dp} \approx \frac{800 - 1175}{2.25 - 2.00} = \frac{-375}{0.25} = -1500$$

so $$E = \left| \frac{p}{q} \cdot \frac{dq}{dp} \right| \approx \left| \frac{2.00}{1175} \cdot (-1500) \right| = 2.55$$

At $p = 2.25$,

$$\frac{dq}{dp} \approx \frac{430 - 800}{2.50 - 2.25} = \frac{-370}{0.25} = -1480$$

so $$E = \left| \frac{p}{q} \cdot \frac{dq}{dp} \right| \approx \left| \frac{2.25}{800} \cdot (-1480) \right| = 4.16$$

Examination of the elasticities for each of the prices suggests that elasticity gets larger as price increases. In other words, at higher prices, an increase in price will cause a larger drop in demand than the same size price increase at a lower price level. This can be explained by the fact that people will not pay too much for candy, as it is somewhat of a "luxury" item.

(c) Elasticity is approximately equal to 1 at $p = \$1.25$ and $p = \$1.50$.

(d)

TABLE 5.1

$p(\$)$	q	Revenue $= p \cdot q \ (\$)$
1.00	2765	2765
1.25	2440	3050
1.50	1980	2970
1.75	1660	2905
2.00	1175	2350
2.25	800	1800
2.50	430	1075

We can see that revenue is maximized at $p = \$1.25$, with $p = \$1.50$ a close second, which agrees with part (c).

17. Demand is elastic at all prices. No matter what the price is, you can increase revenue by lowering the price. In the end, you would lower your prices all the way to zero. This is not a realistic example, but it is mathematically possible. It would correspond, for instance, to the demand equation $q = 1/p^2$, which gives revenue $R = pq = 1/p$ which is decreasing for all prices $p > 0$.

21. The approximation $E_{cross} \approx \left| \frac{\Delta q/q}{\Delta p/p} \right|$ shows that the cross-price elasticity measures the ratio of the fractional change in quantity of chicken demanded to the fractional change in the price of beef. Thus, for example, a 1% increase in the price of beef will stimulate a $E_{cross}\%$ increase in the demand for chicken, presumably because consumers will react to the price rise in beef by switching to chicken. The cross-price elasticity measures the strength of this switch.

Solutions for Section 5.6

1. Substituting $t = 0, 10, 20, \ldots, 70$ into the function $P = 3.9(1.03)^t$ gives the values in Table 5.2. Notice that the agreement is very close, reflecting the fact that an exponential function models the growth well over the period 1790–1860.

TABLE 5.2 *Predicted versus actual US population 1790–1860, in millions. (exponential model)*

Year	Actual	Predicted	Year	Actual	Predicted
1790	3.9	3.9	1830	12.9	12.7
1800	5.3	5.2	1840	17.1	17.1
1810	7.2	7.0	1850	23.2	23.0
1820	9.6	9.5	1860	31.4	30.9

5. (a) If we graph the data, we see that it looks like logistic growth. (See Figure 5.21.) But logistic growth also makes sense from a common-sense viewpoint. As VCRs "catch on," the percentage of households which have them will at first grow exponentially but then slow down after more and more people have them. Eventually, nearly everyone who will buy one already has and the percentage levels off.

Figure 5.21

(b) The point of diminishing returns happens around 36%. This predicts a carrying capacity of 70% which is pretty close to the 71.9% we see in 1990 and 1991.

(c) 75%

(d) The limiting value predicts the percentage of households with a television set that will eventually have VCRs. This model predicts that there will never be a time when more than 75% of the households with TVs also have VCRs, which seems reasonable.

9. (a) We use $k = 1.78$ as a rough approximation. We let $L = 5000$ since the problem tells us that 5000 people eventually get the virus. This means the limiting value is 5000.

    (b) We know that

$$P(t) = \frac{5000}{1 + Ce^{-1.78t}} \quad \text{and} \quad P(0) = 10$$

so

$$10 = \frac{5000}{1 + Ce^0} = \frac{5000}{1 + C}$$
$$10(1 + C) = 5000$$
$$1 + C = 500$$
$$C = 499.$$

    (c) We have $P(t) = \frac{5000}{1 + 499e^{-1.78t}}$. This function is graphed in Figure 5.22.

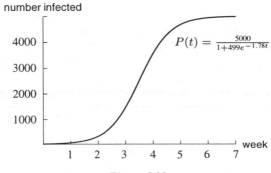

Figure 5.22

    (d) The point of diminishing returns appears to be at the point $(3.5, 2500)$; that is, after 3 and a half weeks and when 2500 people are infected.

13. (a) The dose-response curve for product C crosses the minimum desired response line last, so it requires the largest dose to achieve the desired response. The dose-response curve for product B crosses the minimum desired response line first, so it requires the smallest dose to achieve the desired response.

    (b) The dose-response curve for product A levels off at the highest point, so it has the largest maximum response. The dose-response curve for product B levels off at the lowest point, so it has the smallest maximum response.

    (c) Product C is the safest to administer because its slope in the safe and effective region is the least, so there is a broad range of dosages for which the drug is both safe and effective.

17. When 50 mg of the drug is administered, it is effective for 85 percent of the patients and lethal for 6 percent.

Solutions for Section 5.7

1.

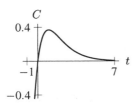

Figure 5.23: $a = 1$

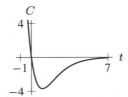

Figure 5.24: $a = -10$

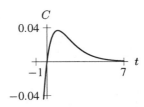

Figure 5.25: $a = 0.1$

The parameter a apparently affects the height and direction of $C = ate^{-bt}$. If a is positive, the "hump" is above the t-axis. If a is negative, it's below the t-axis. If $a > 0$, the larger the value of a, the larger the maximum value of C. If $a < 0$, the more negative the value of a, the smaller the minimum value of C.

5.

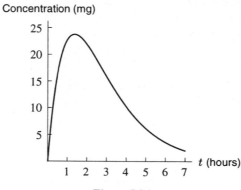

Concentration (mg)

Figure 5.26

9. Food dramatically increases the value of the peak concentration but does not affect the time it takes to reach the peak concentration. The effect of food is stronger during the first 8 hours.

Solutions for Chapter 5 Review

1.

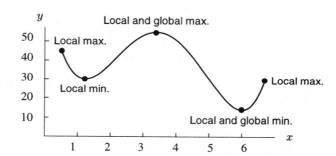

5. (a) Differentiating $f(x) = x + \sin x$ produces $f'(x) = 1 + \cos x$. A second differentiation produces $f''(x) = -\sin x$.
   (b) $f'(x)$ is defined for all x and $f'(x) = 0$ when $x = \pi$. Thus π is the critical point of f.
   (c) $f''(x)$ is defined for all x and $f''(x) = 0$ when $x = 0$, $x = \pi$ and $x = 2\pi$. Since the concavity of f changes at each of these points they are all inflection points.
   (d) $f(0) = 0$, $f(2\pi) = 2\pi$, $f(\pi) = \pi$. So f has a global minimum at $x = 0$ and a global maximum at $x = 2\pi$.
   (e) Plotting the function $f(x)$ for $0 \leq x \leq 2\pi$ gives the graph shown in Figure 5.27:

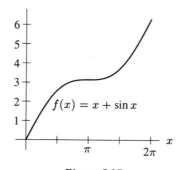

$f(x) = x + \sin x$

Figure 5.27

9. (a) Increasing for $x > 0$, decreasing for $x < 0$.

   (b) $f(0)$ is a local and global minimum, and f has no global maximum.

13. We know that the maximum (or minimum) profit can occur when

$$\text{Marginal cost } = \text{ Marginal revenue } \quad \text{or} \quad MC = MR.$$

From the table it appears that $MC = MR$ at $q \approx 2500$ and $q \approx 4500$. To decide which one corresponds to the maximum profit, look at the marginal profit at these points. Since

$$\text{Marginal profit } = \text{ Marginal revenue } - \text{ Marginal cost}$$

(or $M\pi = MR - MC$), we compute marginal profit at the different values of q in Table 5.3:

TABLE 5.3

q	1000	2000	3000	4000	5000	6000
$M\pi = MR - MC$	−22	−4	4	7	−5	−22

From the table, at $q \approx 2500$, we see that profit changes from decreasing to increasing, so $q \approx 2500$ gives a local minimum. At $q \approx 4500$, profit changes from increasing to decreasing, so $q \approx 4500$ is a local maximum. See Figure 5.28. Therefore, the global maximum occurs at $q = 4500$ or at the endpoint $q = 1000$.

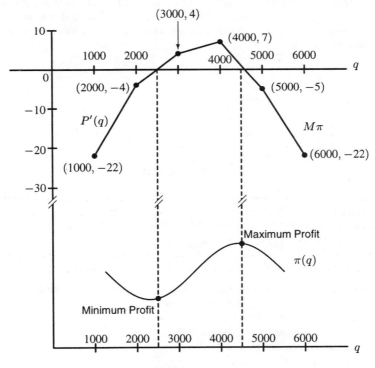

Figure 5.28

17. (a) Profit $= \pi = R - C$; profit is maximized when the slopes of the two graphs are equal, at around $q = 350$. See Figure 5.29.

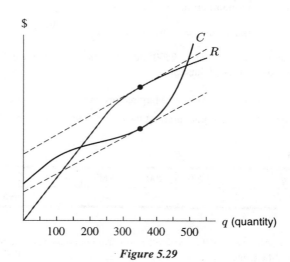

Figure 5.29

(b) The graphs of MR and MC are the derivatives of the graphs of R and C. Both R and C are increasing everywhere, so MR and MC are everywhere positive. The cost function is concave down and then concave up, so MC is decreasing and then increasing. The revenue function is linear and then concave down, so MR is constant and then decreasing. See Figure 5.30.

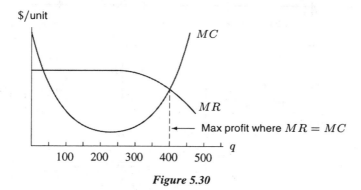

Figure 5.30

21. (a) The IV method reaches peak concentration the fastest, it in fact begins at its peak.
   The P-IM method reaches peak concentration the slowest.
   (b) The IV method has the largest peak concentration.
   The PO method has the smallest peak concentration.
   (c) The IV method wears off the fastest.
   The P-IM method wears off the slowest.
   (d) The P-IM method has the longest effective duration.
   The IV method has the shortest effective duration.
   (e) It is effective for approximately 5 hours.

25. Since marginal revenue equals dR/dq and $R = pq$, we have, using the product rule,

$$\frac{dR}{dq} = \frac{d(pq)}{dq} = p \cdot 1 + \frac{dp}{dq} \cdot q = p\left(1 + \frac{q}{p} \cdot \frac{dp}{dq}\right) = p\left(1 - \frac{1}{-\frac{p}{q} \cdot \frac{dq}{dp}}\right) = p\left(1 - \frac{1}{E}\right).$$

29.

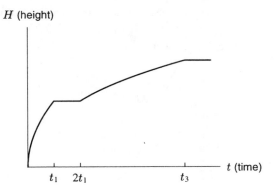

H (height)

t_1 $2t_1$ t_3 t (time)

Suppose t_1 is the time to fill the left side to the top of the middle ridge. Since the container gets wider as you go up, the rate dH/dt decreases with time. Therefore, for $0 \leq t \leq t_1$, graph is concave down.

At $t = t_1$, water starts to spill over to right side and so depth of left side doesn't change. It takes as long for the right side to fill to the ridge as the left side, namely t_1. Thus the graph is horizontal for $t_1 \leq t \leq 2t_1$.

For $t \geq 2t_1$, water level is above the central ridge. The graph is climbing because the depth is increasing, but at a slower rate than for $t \leq t_1$ because the container is wider. The graph is concave down because width is increasing with depth. Time t_3 represents the time when container is full.

33. Recall that the natural logarithm is undefined for $x \leq 0$, so the domain of f is $x > 0$. We see from looking at the graph of $f(x) = x - \ln x$ in the text that this function has one local minimum. We want to assign values to a and b so that this local minimum occurs at $x = 2$. The function must therefore have a critical point at $x = 2$. We find the derivative of $f(x) = a(x - b \ln x)$ and the critical points in terms of a and b.

$$f'(x) = a\left(1 - b\left(\frac{1}{x}\right)\right) = 0$$
$$1 - b\left(\frac{1}{x}\right) = 0$$
$$1 = \frac{b}{x}$$
$$x = b$$

We see that $f(x)$ has only one critical point, at $x = b$. Since we want a critical point at $x = 2$, we choose $b = 2$.

Since $b = 2$, we have $f(x) = a(x - 2\ln x)$. We now use the condition that $f(2) = 5$ to find a:

$$f(2) = 5$$
$$a(2 - 2\ln 2) = 5$$
$$a = 5/(2 - 2\ln 2)$$
$$a \approx 8.147.$$

We let $a = 8.147$ and $b = 2$, so the function is $f(x) = 8.147(x - 2\ln x)$. If we sketch a graph of this function, we see that this function does indeed have a local minimum approximately at the point $(2, 5)$.

37. (a) Differentiating using the product rule gives

$$C'(t) = 20e^{-0.03t} + 20t(-0.03)e^{-0.03t}$$
$$= 20(1 - 0.03t)e^{-0.03t}.$$

At the peak concentration, $C'(t) = 0$, so

$$20(1 - 0.03t)e^{-0.03t} = 0$$
$$t = \frac{1}{0.03} = 33.3 \text{ minutes.}$$

When $t = 33.3$, the concentration is

$$C = 20(33.3)e^{-0.03(33.3)} \approx 245 \text{ ng/ml.}$$

See Figure 5.31. The curve peaks after 33.3 minutes with a concentration of 244.9 ng/ml.

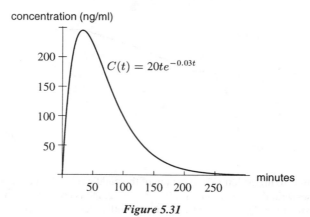

Figure 5.31

(b)   After 15 minutes, the drug concentration will be $C(15) = 20(15)e^{(-0.03)(15)} \approx 191$ ng/ml. After an hour, the concentration will be $C(60) = 20(60)e^{-0.03(60)} \approx 198$ ng/ml.

(c)   We want to know where $C(t) = 10$. We estimate from the graph. It looks like $C(t) = 10$ after 190 minutes or a little over 3 hours.

CHAPTER SIX

Solutions for Section 6.1

1. By counting boxes, we find $\int_1^6 f(x)\,dx = 8.5$, so the average value of f is $\dfrac{8.5}{6-1} = \dfrac{8.5}{5} = 1.7$.

5. Average value = $\dfrac{1}{10-0}\int_0^{10} e^t\,dt \approx 2202.55$

9. (a) The average inventory is given by the formula

$$\frac{1}{90-0}\int_0^{90} 5000(0.9)^t\,dt.$$

Using a calculator yields

$$\int_0^{90} 5000(0.9)^t\,dt \approx 47452.5$$

so the average inventory is

$$\frac{47452.5}{90} \approx 527.25.$$

(b) The function is graphed in Figure 6.1. The area of a rectangle of height 527.25 is equal to the area under the curve.

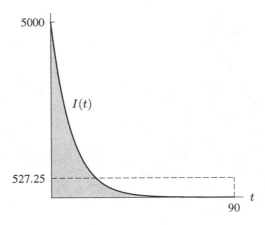

Figure 6.1

13. (a)

$$\text{Average population} = \frac{1}{10}\int_0^{10} 67.38(1.026)^t\,dt$$

Evaluating the integral numerically gives

$$\text{Average population} \approx 76.8 \text{ million}$$

(b) In 1980, $t = 0$, and $P = 67.38(1.026)^0 = 67.38$.
In 1990, $t = 10$, and $P = 67.38(1.026)^{10} = 87.10$.
Average= $\frac{1}{2}(67.38 + 87.10) = 77.24$ million.

(c) If P had been linear, the average value found in (a) would have been the one we found in (b). Since the population graph is concave up, it is below the secant line. Thus, the actual values of P are less than the corresponding values on the secant line, and so the average found in (a) is smaller than that in (b).

17. In (a), $f'(1)$ is the slope of a tangent line at $x = 1$, which is negative. As for (c), the rate of change in $f(x)$ is given by $f'(x)$, and the average value of this over $0 \leq x \leq a$ is

$$\frac{1}{a - 0} \int_0^a f'(x)\, dx = \frac{f(a) - f(0)}{a - 0}.$$

This is the slope of the line through the points $(0, 1)$ and $(a, 0)$, which is less negative that the tangent line at $x = 1$. Therefore, $(a) < (c) < 0$. The quantity (b) is $\left(\int_0^a f(x)\, dx \right) / a$ and (d) is $\int_0^a f(x)\, dx$, which is the net area under the graph of f (counting the area as negative for f below the x-axis). Since $a > 1$ and $\int_0^a f(x)\, dx > 0$, we have $0 <$(b)$<$(d). Therefore

$$(a) < (c) < (b) < (d).$$

Solutions for Section 6.2

1. (a) Looking at the figure in the problem we see that the equilibrium price is roughly $30 giving an equilibrium quantity of 125 units.

   (b) Consumer surplus is the area above p^* and below the demand curve. Graphically this is represented by the shaded area in Figure 6.2. From the graph we can estimate the shaded area to be roughly 14 squares where each square represents ($25/unit)·(10 units). Thus the consumer surplus is approximately

   $$14 \cdot \$250 = \$3500.$$

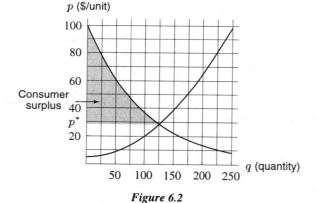

Figure 6.2   **Figure 6.3**

Producer surplus is the area under p^* and above the supply curve. Graphically this is represented by the shaded area in Figure 6.3. From the graph we can estimate the shaded area to be roughly 8 squares where each square represents ($25/unit)·(10 units). Thus the producer surplus is approximately

$$8 \cdot \$250 = \$2000$$

   (c) We have

$$\text{Total gains from trade} = \text{Consumer surplus} + \text{producer surplus}$$
$$= \$3500 + \$2000$$
$$= \$5500.$$

5.

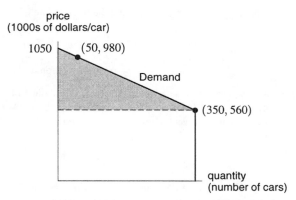

Measuring money in thousands of dollars, the equation of the line representing the demand curve passes through (50, 980) and (350, 560). So the equation is $y - 560 = \frac{420}{-300}(x - 350)$, i.e. $y - 560 = -\frac{7}{5}x + 490$.
The consumer surplus is thus

$$\int_0^{350} \left(-\frac{7}{5}x + 1050\right) dx - (350)(560) = 85750.$$

(Note that $85750 = \frac{1}{2} \cdot 490 \cdot 350$, the area of the triangle in the diagram. We thus could have avoided the formula for consumer surplus in solving the problem.)
Recalling that our unit measure for the price axis is \$1000/car, the consumer surplus is \$85,750,000.

9.

$$\int_0^{q^*} (p^* - S(q))\, dq = \int_0^{q^*} p^*\, dq - \int_0^{q^*} S(q)\, dq$$

$$= p^* q^* - \int_0^{q^*} S(q)\, dq.$$

Using Problem 8, this integral is the extra amount consumers pay (i.e., suppliers earn over and above the minimum they would be willing to accept for supplying the good). It results from charging the equilibrium price.

Solutions for Section 6.3

1.

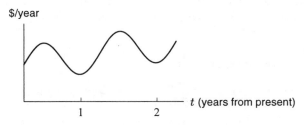

The graph reaches a peak each summer, and a trough each winter. The graph shows sunscreen sales increasing from cycle to cycle. This gradual increase may be due in part to inflation and to population growth.

5. (a) We first find the present value, P, of the income stream:

$$P = \int_0^{10} 6000 e^{-0.05t}\, dt = \$47,216.32.$$

We use the present value to find the future value, F:

$$F = Pe^{rt} = 47126.32 e^{0.05(10)} = \$77,846.55.$$

(b) The income stream contributed \$6000 per year for 10 years, or \$60,000. The interest earned was $77,846.55 - 60,000 = \$17,846.55$.

9.   (a)   Since the income stream is $7035 million per year and the interest rate is 8.5%,

$$\text{Present value} = \int_0^1 7035e^{-0.085t} \, dt$$
$$= 6744.31 \text{ million dollars.}$$

The present value of Intel's profits over a one-year time period is about 6744 million dollars.

(b)   The value at the end of one year is $6744.31e^{0.085(1)} = 7342.64$, or about 7343, million dollars. This is the value, at the end of one year, of Intel's profits over a one-year time period.

13.   Find the time T at which

$$130,000 = \int_0^T 80,000e^{-0.085t} \, dt.$$

Trying a few values of T, we get $T \approx 1.75$. It takes approximately one year and nine months for the present value of the profit generated by the new machinery to equal the cost of the machinery.

Solutions for Section 6.4

1.   (a)   Absolute increase between 1988 and 1989 $= 1275 - 813 = 462$ thousand. Between 1992 and 1993, absolute increase $= 4820 - 3657 = 1163$ thousand.

(b)   Relative increase between 1988 and 1989 $= \frac{462}{813} \approx 56.8\%$. Relative increase between 1992 and 1993 $= \frac{1163}{3657} \approx 31.8\%$.

5.   The relative birth rate is $\frac{30}{1000} = 0.03$ and the relative death rate is $\frac{20}{1000} = 0.02$. The relative rate of growth is $0.03 - 0.02 = 0.01$

9.   The relative growth rate is a constant $0.02 = 2\%$. The change in $\ln P(t)$ is the area under the curve which is $10(0.02) = 0.2$. So

$$\ln P(10) - \ln P(0) = \int_0^{10} \frac{P'(t)}{P(t)} \, dt = 0.2$$
$$\ln\left(\frac{P(10)}{P(0)}\right) = 0.2$$
$$\frac{P(10)}{P(0)} = e^{0.2} \approx 1.22.$$

The population has increased by about 22% over the 10-year period.

Another way of looking at this problem is to say that since $P(t)$ is growing at a constant 2% rate, it is growing exponentially, so

$$P(t) = P(0)e^{0.02t}.$$

Substituting $t = 10$ gives the same result as before:

$$\frac{P(10)}{P(0)} = e^{0.02(10)} = e^{0.2} \approx 1.22.$$

13.   Although the relative growth rate is decreasing, it is everywhere positive, so f is an increasing function for $0 \le t \le 10$.

Solutions for Section 6.5

1. $5x$

5. $\dfrac{x^5}{5}$.

9. $\frac{2}{3}z^{\frac{3}{2}}$

13. $t^3 + \dfrac{7t^2}{2} + t$.

17. $x^3 + 5x$.

21. $\dfrac{y^5}{5} + \ln|y|$

25. $f(x) = 3$, so $F(x) = 3x + C$. $F(0) = 0$ implies that $3 \cdot 0 + C = 0$, so $C = 0$. Thus $F(x) = 3x$ is the only possibility.

29. $f(x) = x^2$, so $F(x) = \dfrac{x^3}{3} + C$. $F(0) = 0$ implies that $\dfrac{0^3}{3} + C = 0$, so $C = 0$. Thus $F(x) = \dfrac{x^3}{3}$ is the only possibility.

33. Since $\dfrac{d}{dx}(e^x) = e^x$, we take $F(x) = e^x + C$. Now

$$F(0) = e^0 + C = 1 + C = 0,$$

so

$$C = -1$$

and

$$F(x) = e^x - 1.$$

37. $2x^3 + C$.

41. $\dfrac{x^4}{4} + 2x^2 + 8x + C$.

45. $\dfrac{x^3}{3} + \ln|x| + C$.

Solutions for Section 6.6

1. If $F'(t) = t^3$, then $F(t) = \dfrac{t^4}{4}$. By the Fundamental Theorem, we have

$$\int_0^3 t^3 \, dt = F(3) - F(0) = \left. \frac{t^4}{4} \right|_0^3 = \frac{3^4}{4} - \frac{0}{4} = \frac{81}{4}.$$

5. Since $F'(y) = y^2 + y^4$, we take $F(y) = \dfrac{y^3}{3} + \dfrac{y^5}{5}$. Then

$$\int_0^1 (y^2 + y^4) \, dy = F(3) - F(0)$$

$$= \left(\frac{1^3}{3} + \frac{1^5}{5} \right) - \left(\frac{0^3}{3} + \frac{0^5}{5} \right)$$

$$= \frac{1}{3} + \frac{1}{5} = \frac{8}{15}.$$

9. $\displaystyle\int_{-3}^{-1} \frac{2}{r^3} \, dr = \left. -r^{-2} \right|_{-3}^{-1} = -1 + \frac{1}{9} = -8/9 \approx -0.889$.

13. $\displaystyle\int_0^1 \sin\theta \, d\theta = \left. -\cos\theta \right|_0^1 = 1 - \cos 1 \approx 0.460$.

17. Figure 6.4 shows the graph of the function $f(t) = e^{-0.2t}$ on the interval $0 \leq t \leq 1$. To make a rough estimate of the area under the curve we will do two things. First, $f(1) \approx 0.82$, so we look at the rectangle bounded by the t- and y-axes, the line $y = 0.82$, and the line $t = 1$. The area of this rectangle is $1 \times (0.82) = 0.82$. We can also observe that the graph of the function closely approximates a straight line on the interval $0 \leq t \leq 1$. Therefore, we have a triangle bounded by the y-axis, the line $y = 0.82$, and the function $f(t) = e^{-0.2t}$, which is approximately a straight line. The area of this triangle is $\frac{1}{2}(1 - .82)(1) = \frac{1}{2}(.18) = 0.09$. Our approximation for the total area is $0.82 + 0.09 = 0.91$. (In Problem 11 we obtained a result of ≈ 0.91 when we evaluated the integral, so our approximation here is quite accurate.)

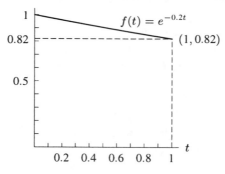

Figure 6.4

21. (a) The definite integral which would measure the total quantity of oil consumed would be

$$\int_0^5 (32e^{0.05t})\, dt.$$

(b) The Fundamental Theorem of Calculus states that

$$\int_a^b f(t)\, dt = F(b) - F(a)$$

provided that $F'(t) = f(t)$. To apply this, we need to find $F(t)$ such that $F'(t) = 32e^{0.05t}$. The function $F(t) = \frac{32}{0.05}e^{0.05t} = 640e^{0.05t}$ will satisfy this requirement (since $\frac{d}{dt}\left(\frac{32}{0.05}e^{0.05t}\right) = (0.05)\frac{32}{0.05}e^{0.05t} = 32e^{0.05t}$). Therefore, the total amount of oil consumed equals

$$\int_0^5 (32e^{0.05t})\, dt = F(5) - F(0) = 640e^{0.05t}\Big|_0^5 = 640(e^{0.25} - e^0) \approx 182.$$

Thus, approximately 182 billion barrels of oil were consumed between 1990 and 1995.

25. (a) Evaluating the integrals with a calculator gives

$$\int_0^{10} xe^{-x/10}\, dx = 26.42$$

$$\int_0^{50} xe^{-x/10}\, dx = 95.96$$

$$\int_0^{100} xe^{-x/10}\, dx = 99.95$$

$$\int_0^{200} xe^{-x/10}\, dx = 100.00$$

(b) The results of part (a) suggest that

$$\int_0^{\infty} xe^{-x/10}\, dx \approx 100$$

29. (a) No, it is not reached since

$$\text{Total number of rabbits} = \int_1^\infty \frac{1}{t^2}\, dt = 1.$$

Thus, the total number of rabbits is 1000.

(b) Yes, since $\int_1^\infty t\, dt$ does not converge to a finite value, which means that infinitely many rabbits could be produced, and therefore 1 million is certainly reached.

(c) Yes, since $\int_1^\infty \frac{1}{\sqrt{t}}\, dt$ does not converge to a finite value.

Solutions for Section 6.7

1. Since $F(0) = 0$, $F(b) = \int_0^b f(t)\, dt$. For each b we determine $F(b)$ graphically as follows:
$F(0) = 0$
$F(1) = F(0) + \text{Area of } 1 \times 1 \text{ rectangle} = 0 + 1 = 1$
$F(2) = F(1) + \text{Area of triangle } (\frac{1}{2} \cdot 1 \cdot 1) = 1 + 0.5 = 1.5$
$F(3) = F(2) + \text{Negative of area of triangle} = 1.5 - 0.5 = 1$
$F(4) = F(3) + \text{Negative of area of rectangle} = 1 - 1 = 0$
$F(5) = F(4) + \text{Negative of area of rectangle} = 0 - 1 = -1$
$F(6) = F(5) + \text{Negative of area of triangle} = -1 - 0.5 = -1.5$
The graph of $F(t)$, for $0 \le t \le 6$, is shown in Figure 6.5.

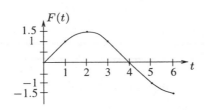

Figure 6.5

5. We see that
F decreases when $x < 1.5$ or $x > 4.67$, because F' is negative there.
F increases when $1.5 < x < 4.67$, because F' is positive there.
So
F has a local minimum at $x = 1.5$.
F has a local maximum at $x = 4.67$.
We have $F(0) = 14$. Since F' is negative between 0 and 1.5, the Fundamental Theorem of Calculus gives us

$$F(1.5) - F(0) = \int_0^{1.5} F'(x)\, dx = -34$$
$$F(1.5) = 14 - 34 = -20.$$

Similarly

$$F(4.67) = F(1.5) + \int_{1.5}^{4.67} F'(x)\, dx = -20 + 25 = 5.$$
$$F(6) = F(4.67) + \int_{4.67}^{6} F'(x)\, dx = 5 - 5 = 0.$$

A graph of F is in Figure 6.6. The local maximum is $(4.67, 5)$ and the local minimum is $(1.5, -20)$.

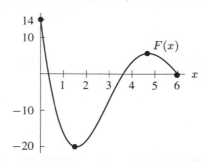

Figure 6.6

9. By the Fundamental Theorem,

$$f(1) - f(0) = \int_0^1 f'(x)\,dx,$$

Since $f'(x)$ is negative for $0 \le x \le 1$, this integral must be negative and so $f(1) < f(0)$.

13.

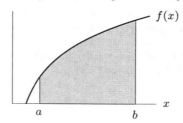

Solutions for Section 6.8

1. Since $p(x)$ is a density function, the area under the graph of $p(x)$ is 1, so

$$\text{Area} = \text{Base} \; \cdot \; \text{Height} = 15a = 1$$

$$a = \frac{1}{15}.$$

5. Since $p(t)$ is a density function,

$$\text{Area under graph} = \frac{1}{2} \cdot c \cdot 0.01 = 0.005c = 1,$$

so $c = 1/0.005 = 200$.

9. We use the fact that the area of a triangle is $\frac{1}{2} \cdot \text{Base} \cdot \text{Height}$. Since $p(x)$ is a line with slope $0.1/20 = 0.005$, its equation is

$$p(x) = 0.005x.$$

(a) The fraction less than 5 meters high is the area to the left of 5. Since $p(5) = 0.005(5) = 0.025$,

$$\text{Fraction} = \frac{1}{2} \cdot 5(0.025) = 0.0625.$$

(b) The fraction more than 6 meters high is the area to the right of 6. Since $p(6) = 0.005(6) = 0.03$,

$$\text{Fraction} = 1 - (\text{Area to left of } 6)$$

$$= 1 - \frac{1}{2} \cdot 6(0.03) = 0.91.$$

(c) Fraction between 2 and 5 meters high is area between 2 and 5. Since $p(2) = 0.005(2) = 0.01$,

$$\text{Fraction} = (\text{Area to left of } 5) - (\text{Area to left of } 2)$$

$$= 0.0625 - \frac{1}{2} \cdot 2 \cdot (0.01) = 0.0525.$$

13.   See Figure 6.7. Many other answers are possible.

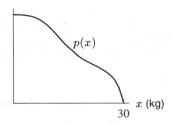

Figure 6.7

17.   No. Though the density function has its maximum value at 50, this does not mean that a large fraction of the population receives scores near 50. The value $p(50)$ can not be interpreted as a probability. Probability corresponds to *area* under the graph of a density function. Most of the area in this case is in the broad hump covering the range $0 \leq x \leq 40$, very little in the peak around $x = 50$. Most people score in the range $0 \leq x \leq 40$.

Solutions for Section 6.9

1.

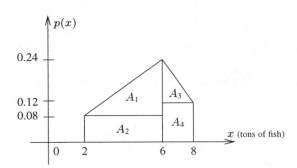

Splitting the figure into four pieces, we see that

$$\text{Area under the curve} = A_1 + A_2 + A_3 + A_4$$
$$= \frac{1}{2}(0.16)4 + 4(0.08) + \frac{1}{2}(0.12)2 + 2(0.12)$$
$$= 1.$$

We expect the area to be 1, since $\int_{-\infty}^{\infty} p(x)\, dx = 1$ for any probability density function, and $p(x)$ is 0 except when $2 \leq x \leq 8$.

5.

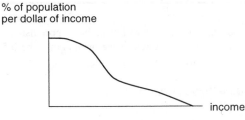

Figure 6.8: Density function

Figure 6.9: Cumulative distribution function

9.   (a)   The first item is sold at the point at which the graph is first greater than zero. Thus the first item is sold at $t = 30$ or January 30. The last item is sold at the t value at which the function is first equal to 100%. Thus the last item is sold at $t = 240$ or August 28, unless its a leap year.

   (b)   Looking at the graph at $t = 121$ we see that roughly 65% of the inventory has been sold by May 1.

   (c)   The percent of the inventory sold during May and June is the difference between the percent of the inventory sold on the last day of June and the percent of the inventory sold on the first day of May. Thus, the percent of the inventory sold during May and June is roughly 25%.

   (d)   The percent of the inventory left after half a year is

$$100 - \text{(percent inventory sold after half year)}.$$

   Thus, roughly 10% of the inventory is left after half a year.

   (e)   The items probably went on sale on day 100 and were on sale until day 120. Roughly from April 10 until April 30.

13. (a)   The fraction of students passing is given by the area under the curve from 2 to 4 divided by the total area under the curve. This appears to be about $\frac{2}{3}$.

   (b)   The fraction with honor grades corresponds to the area under the curve from 3 to 4 divided by the total area. This is about $\frac{1}{3}$.

   (c)   The peak around 2 probably exists because many students work to get just a passing grade.

   (d)

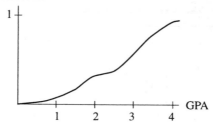

17.   Figure 6.10 is a graph of the density function; Figure 6.11 is a graph of the cumulative distribution.

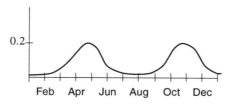

Figure 6.10: Density function

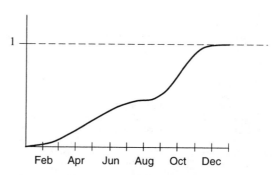

Figure 6.11: Cumulative distribution function

Solutions for Section 6.10

1.   The median daily catch is the amount of fish such that half the time a boat will bring back more fish and half the time a boat will bring back less fish. Thus the area under the curve and to the left of the median must be 0.5. There are 25 squares under the curve so the median occurs at 12.5 squares of area. Now

$$\int_2^5 p(x)\,dx = 10.5 \text{ squares}$$

and

$$\int_5^6 p(x)\,dx = 5.5 \text{ squares},$$

so the median occurs at a little over 5 tons. We must find the value a for which

$$\int_5^a p(t)dt = 2 \text{ squares},$$

and we note that this occurs at about $a = 0.35$. Hence

$$\int_2^{5.35} p(t)\, dt \approx 12.5 \text{ squares}$$

$$\approx 0.5.$$

The median is about 5.35 tons.

5.   We know that the mean is given by

$$\int_{-\infty}^{\infty} tp(t)dt.$$

Thus we get

$$\text{Mean } = \int_0^4 tp(t)dt$$

$$= \int_0^4 (-0.0375t^3 + 0.225t^2)dt$$

$$\approx 2.4$$

Thus the mean is 2.4 weeks. Figure 6.12 supports the conclusion that $t = 2.4$ is in fact the mean.

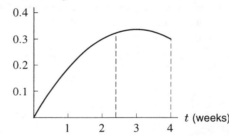

Figure 6.12

9.   (a) P is the cumulative distribution function, so the percentage of the population that made between \$20,000 and \$50,000 is

$$P(50) - P(20) = 99\% - 75\% = 24\%.$$

Therefore $\frac{6}{25}$ of the population made between \$20,000 and \$50,000.

(b)   The median income is the income such that half the people made less than this amount. Looking at the chart, we see that $P(12.6) = 50\%$, so the median must be \$12,600.

(c)   The cumulative distribution function looks something like this:

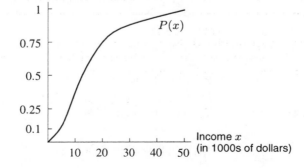

The density function is the derivative of the cumulative distribution. Qualitatively it looks like:

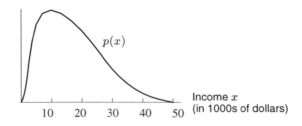

The density function has a maximum at about $8000. This means that more people have incomes around $8000 than around any other amount. On the density function, this is the highest point. On the cumulative distribution, this is the point of steepest slope (because $P' = p$), which is also the point of inflection.

Solutions for Chapter 6 Review

1. The average value equals

$$\frac{1}{3} \int_0^3 f(x)\, dx = \frac{24}{3} = 8.$$

5. Looking at the graph we see that the supply and demand curves intersect at roughly the point $(345, 8)$. Thus the equilibrium price is $8 per unit and the equilibrium quantity is 345 units. Figures 6.13 and 6.14 show the shaded areas corresponding to the consumer surplus and the producer surplus.

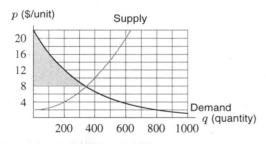

Figure 6.13: Consumer surplus

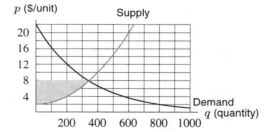

Figure 6.14: Producer surplus

Counting boxes we see that the consumer surplus is roughly $2000 while the producer surplus is roughly $1400.

9. (a) If P is the present value, then the value in two years at 9% interest is $Pe^{0.09(2)}$:

$$500,000 = Pe^{0.09(2)}$$
$$P = \frac{500,000}{e^{0.09(2)}} = 417,635.11$$

The present value of the renovations is $417,635.11.

(b) If money is invested at a constant rate of $S per year, then

$$\text{Present value of deposits} = \int_0^2 Se^{-0.09t}\, dt$$

Since S is constant, we can take it out in front of the integral sign:

$$\text{Present value of deposits} = S \int_0^2 e^{-0.09t}\, dt$$

We want the rate S so that the present value is 417,635.11:

$$417{,}635.11 = S \int_0^2 e^{-0.09t}\, dt$$

$$417{,}635.11 = S(1.830330984)$$

$$S = \frac{417{,}635.11}{1.830330984} = 228{,}174.64.$$

Money deposited at a continuous rate of 228,174.64 dollars per year and earning 9% interest per year has a value of \$500,000 after two years.

13. $\ln|x| - \dfrac{1}{x} - \dfrac{1}{2x^2} + C$

17. $G(\theta) = -\cos\theta - 2\sin\theta + C$

21. $\dfrac{t^3}{3} - 3t^2 + 5t + C.$

25. $3\ln|t| + \dfrac{2}{t} + C$

29. $\dfrac{x^2}{2} + 2\ln|x| - \pi\cos x + C$

33.   If $f(t) = e^{0.05t}$, then $F(t) = 20e^{0.05t}$ (you can check this by observing that $\dfrac{d}{dt}(20e^{0.05t}) = e^{0.05t}$). By the Fundamental Theorem, we have

$$\int e^{0.05t}\, dt = 20e^{0.05t}\Big|_0^3 = 20e^{0.15} - 20e^0 = 20(e^{0.15} - 1).$$

37.   Suppose x is the age of death; a possible density function is graphed in Figure 6.15.

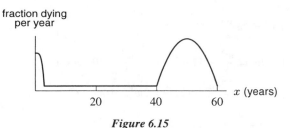

Figure 6.15

41.   (a)   Using a calculator or computer, we get

$$\int_0^3 e^{-2t}\, dt = 0.4988 \qquad\qquad \int_0^5 e^{-2t}\, dt = 0.49998$$

$$\int_1^7 e^{-2t}\, dt = 0.4999996 \qquad\qquad \int_0^{10} e^{-2t}\, dt = 0.499999999.$$

The values of these integrals are getting closer to 0.5. A reasonable guess is that the improper integral converges to 0.5.

   (b)   Since $-\tfrac{1}{2}e^{-2t}$ is an antiderivative of e^{-2t}, we have

$$\int_0^b e^{-2t}\, dt = -\frac{1}{2}e^{-2t}\Big|_0^b = -\frac{1}{2}e^{-2b} - \left(-\frac{1}{2}e^0\right) = -\frac{1}{2}e^{-2b} + \frac{1}{2}.$$

   (c)   Since $e^{-2b} = 1/e^{2b}$, we have

$$e^{2b} \to \infty \quad \text{as} \quad b \to \infty, \quad \text{so} \quad e^{-2b} = \frac{1}{e^{2b}} \to 0.$$

Therefore,

$$\lim_{b \to \infty} \int_0^b e^{-2t}\, dt = \lim_{b \to \infty} \left(-\frac{1}{2} e^{-2b} + \frac{1}{2} \right) = 0 + \frac{1}{2} = \frac{1}{2}.$$

So the improper integral converges to $1/2 = 0.5$:

$$\int_0^\infty e^{-2t}\, dt = \frac{1}{2}.$$

45. (a) The probability you dropped the glove within a kilometer of home is given by

$$\int_0^1 2e^{-2x}\, dx = \left. -e^{-2x} \right|_0^1 = -e^{-2} + 1 \approx 0.865.$$

(b) Since the probability that the glove was dropped within y km $= \int_0^y p(x)\,dx = 1 - e^{-2y}$, we solve

$$1 - e^{-2y} = 0.95$$
$$e^{-2y} = 0.05$$
$$y = \frac{\ln 0.05}{-2} \approx 1.5 \text{ km.}$$

Solutions to Practice Problems on Integration

1. $\displaystyle \int (t^3 + 6t^2)\, dt = \frac{t^4}{4} + 6 \cdot \frac{t^3}{3} + C = \frac{t^4}{4} + 2t^3 + C$

5. $\displaystyle \int 3w^{1/2}\, dw = 3 \cdot \frac{w^{3/2}}{3/2} + C = 2w^{3/2} + C$

9. $\displaystyle \int (w^4 - 12w^3 + 6w^2 - 10)\, dw = \frac{w^5}{5} - 12 \cdot \frac{w^4}{4} + 6 \cdot \frac{w^3}{3} - 10 \cdot w + C$

$$= \frac{w^5}{5} - 3w^4 + 2w^3 - 10w + C$$

13. $\displaystyle \int \left(\frac{4}{x} + 5x^{-2} \right)\, dx = 4 \ln |x| + \frac{5x^{-1}}{-1} + C = 4 \ln |x| - \frac{5}{x} + C$

17. $\displaystyle \int (5 \sin x + 3 \cos x)\, dx = -5 \cos x + 3 \sin x + C$

21. $\displaystyle \int 15p^2 q^4\, dp = 15 \left(\frac{p^3}{3} \right) q^4 + C = 5p^3 q^4 + C$

25. $\displaystyle \int 5e^{2q}\, dq = 5 \cdot \frac{1}{2} e^{2q} + C = 2.5e^{2q} + C$

29. $\displaystyle \int (x^2 + 8 + e^x)\, dx = \frac{x^3}{3} + 8x + e^x + C$

CHAPTER SEVEN

Solutions for Section 7.1

1. (a) Beef consumption by households making \$20,000/year is given by Row 1 of Table 7.2 on page 358 of the text.

TABLE 7.1

p	3.00	3.50	4.00	4.50
$f(20, p)$	2.65	2.59	2.51	2.43

For households making \$20,000/year, beef consumption decreases as price goes up.

(b) Beef consumption by households making \$100, 000/year is given by Row 5 of Table 7.2.

TABLE 7.2

p	3.00	3.50	4.00	4.50
$f(100, p)$	5.79	5.77	5.60	5.53

For households making \$100,000/year, beef consumption also decreases as price goes up.

(c) Beef consumption by households when the price of beef is \$3.00/lb is given by Column 1 of Table 7.2.

TABLE 7.3

I	20	40	60	80	100
$f(I, 3.00)$	2.65	4.14	5.11	5.35	5.79

When the price of beef is \$3.00/lb, beef consumption increases as income increases.

(d) Beef consumption by households when the price of beef is \$4.00/lb is given by Column 3 of Table 7.2.

TABLE 7.4

I	20	40	60	80	100
$f(I, 4.00)$	2.51	3.94	4.97	5.19	5.60

When the price of beef is \$4.00/lb, beef consumption increases as income increases.

5. In the answer to Problem 4 we saw that

$$P = 0.052 \frac{M}{I},$$

and in the answer to Problem 3 we saw that

$$M = pf(I, p).$$

Putting the expression for M into the expression for P, gives:

$$P = 0.052 \frac{pf(I, p)}{I}.$$

9. To see whether f is an increasing or decreasing function of x, we need to see how f varies as we increase x and hold y fixed. We note that each column of the table corresponds to a fixed value of y. Scanning down the $y = 2$ column, we can see that as x increases, the value of the function decreases from 114 when $x = 0$ down to 93 when $x = 80$. Thus, f may be decreasing. In order for f to actually be decreasing however, we have to make sure that f decreases for *every* column. In this case, we see that f indeed does decrease for every column. Thus, f is a decreasing function of x. Similarly, to see whether f is a decreasing function of y we need to look at the rows of the table. As we can see, f increases for every row as we increase y. Thus, f is an increasing function of y.

13. (a) According to Table 7.4 of the problem, it feels like $-31°$F.
    (b) A wind of 10 mph, according to Table 7.4.
    (c) About 5.5 mph. Since at a temperature of $25°$F, when the wind increases from 5 mph to 10 mph, the temperature adjusted for wind-chill decreases from $21°$F to $10°$F, we can say that a 5 mph increase in wind speed causes an $11°$F decrease in the temperature adjusted for wind-chill. Thus, each 0.5 mph increase in wind speed brings *about* a $1°$F drop in the temperature adjusted for wind-chill.
    (d) With a wind of 15 mph, approximately $23.5°$F would feel like $0°$F. With a 15 mph wind speed, when air temperature drops five degrees from $25°$F to $20°$F, the temperature adjusted for wind-chill drops 7 degrees from $2°$F to $-5°$F. We can say that for every $1°$F decrease in temperature there is *about* a $1.4°$F $(= 7/5)$ drop in the temperature you feel.

17.

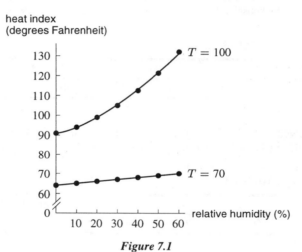

Figure 7.1

Both graphs are increasing because at any fixed temperature the air feels hotter as the humidity increases. The fact that the graph for $T = 100$ increases more rapidly with humidity than the graph for $T = 70$ tells us that when it is hot $(100°$F), high humidity has more effect on how we feel than at lower temperatures $(70°$F).

21. It stands to reason that the demand for tea, D, will have a similar formula to the demand for coffee. The demand D will be a decreasing function of the price of tea, t, and an increasing function of the price of coffee c. A possible formula for the demand for tea is

$$D = 100\frac{c}{t}.$$

25. (a) For $g(x,t) = \cos 2t \sin x$, our snapshots for fixed values of t are still one arch of the sine curve. The amplitudes, which are governed by the $\cos 2t$ factor, now change twice as fast as before. That is, the string is vibrating twice as fast.
    (b) For $y = h(x,t) = \cos t \sin 2x$, the vibration of the string is more complicated. If we hold t fixed at any value, the snapshot now shows one full period, i.e. one crest and one trough, of the sine curve. The magnitude of the sine curve is time dependent, given by $\cos t$. Now the center of the string, $x = \pi/2$, remains stationary just like the end points. This is a vibrating string with the center held fixed, as shown in Figure 7.2.

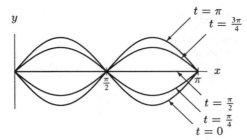

Figure 7.2: Another vibrating string:
$$y = h(x,t) = \cos t \sin 2x$$

Solutions for Section 7.2

1. (a) 80-90°F
   (b) 60-72°F
   (c) 60-100°F

5. We can see that as we move horizontally to the right, we are increasing x but not changing y. As we take such a path at $y = 2$, we cross decreasing contour lines, starting at the contour line 6 at $x = 1$ to the contour line 1 at around $x = 5.7$. This trend holds true for all of horizontal paths. Thus, z is a decreasing function of x. Similarly, as we move up along a vertical line, we cross increasing contour lines and thus z is an increasing function of y.

9. To draw a contour for a wind-chill of $W = 20$, we need a few combinations of temperature and wind velocity (T, v) such that $W(T, v) = 20$. Estimating from the table, some such points are $(24, 5)$ and $(33, 10)$. We can connect these points to get a contour for $W = 20$. Similarly, some points that have wind-chill of about $0°F$ are $(5, 5)$, $(17.5, 10)$, $(23.5, 15)$, $(27, 20)$, and $(29, 25)$. By connecting these points we get the contour for $W = 0$. If we carry out this procedure for more values of W, we get a full contour diagram such as is shown in Figure 7.3:

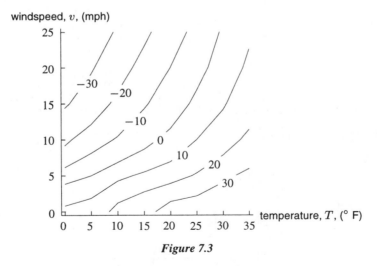

Figure 7.3

13.   (a)   Graph with kilometers north fixed at 50:

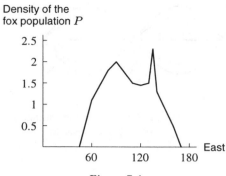

Figure 7.4

(b)   Graph with kilometers north fixed at 100:

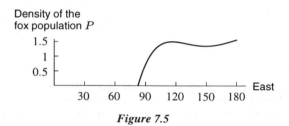

Figure 7.5

(c)   Graph with kilometers east fixed at 60:

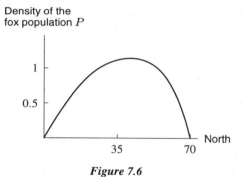

Figure 7.6

(d)   Graph with kilometers east fixed at 120:

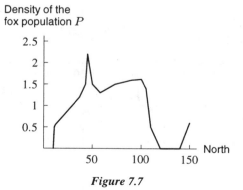

Figure 7.7

17.   The contour where $f(x, y) = -x - y = c$ or $y = -x - c$ is the graph of the straight line of slope -1 as shown in Figure 7.8. Note that we have plotted contours for $c = -3, -2, -1, 0, 1, 2, 3$.

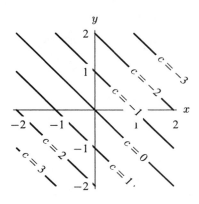

Figure 7.8

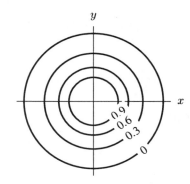

Figure 7.9

21.   A possible contour diagram showing light intensity in the park as a function of position is in Figure 7.9. We can either label the contours as in Figure 7.9, or in reverse order. Figure 7.9 assumes that the park is well lit, and that the coastline is dark. Alternatively, we could assume that the park is dark at night, but that near the coastline are a lot of tourist areas, maybe a carnival, or a clambake and bonfire. Many answers are possible.

25.   (a)   The contour lines are much closer together on path A, so path A is steeper.
      (b)   If you are on path A and turn around to look at the countryside, you find hills to your left and right, obscuring the view. But the ground falls away on either side of path B, so you are likely to get a much better view of the countryside from path B.
      (c)   There is more likely to be a stream alongside path A, because water follows the direction of steepest descent.

29.   The point $x = 10$, $t = 5$ is between the contours $H = 70$ and $H = 75$, a little closer to the former. Therefore, we estimate $H(10, 5) \approx 72$, i.e., it is about $72°$F. Five minutes later we are at the point $x = 10$, $t = 10$, which is just above the contour $H = 75$, so we estimate that it has warmed up to $76°$F by then.

Solutions for Section 7.3

1.   (a)   We expect the demand for coffee to decrease as the price of coffee increases (assuming the price of tea is fixed.) Thus we expect f_c to be negative. We expect people to switch to coffee as the price of tea increases (assuming the price of coffee is fixed), so that the demand for coffee will increase. We expect f_t to be positive.
      (b)   The statement $f(3, 2) = 780$ tells us that if coffee costs \$3 per pound and tea costs \$2 per pound, we can expect 780 pounds of coffee to sell each week. The statement $f_c(3, 2) = -60$ tells us that, if the price of coffee then goes up \$1 and the price of tea stays the same, the demand for coffee will go down by about 60 pounds. The statement $20 = f_t(3, 2)$ tells us that if the price of tea goes up \$1 and the price of coffee stays the same, the demand for coffee will go up by about 20 pounds.

5. $\partial P / \partial t$: The unit is dollars per month. This is the rate at which payments change as the number of months it takes to pay off the loan changes. The sign is negative because payments decrease as the pay-off time increases.

      $\partial P / \partial r$: The unit is dollars per percentage point. This is the rate at which payments change as the interest rate changes. The sign is positive because payments increase as the interest rate increases.

9.   If h is small, then

$$f_x(3, 2) \approx \frac{f(3 + h, 2) - f(3, 2)}{h}.$$

With $h = 0.01$, we find

$$f_x(3, 2) \approx \frac{f(3.01, 2) - f(3, 2)}{0.01} = \frac{\frac{3.01^2}{(2+1)} - \frac{3^2}{(2+1)}}{0.01} = 2.00333.$$

With $h = 0.0001$, we get

$$f_x(3, 2) \approx \frac{f(3.0001, 2) - f(3, 2)}{0.0001} = \frac{\frac{3.0001^2}{(2+1)} - \frac{3^2}{(2+1)}}{0.0001} = 2.0000333.$$

Since the difference quotient seems to be approaching 2 as h gets smaller, we conclude

$$f_x(3, 2) \approx 2.$$

To estimate $f_y(3, 2)$, we use

$$f_y(3, 2) \approx \frac{f(3, 2+h) - f(3, 2)}{h}.$$

With $h = 0.01$, we get

$$f_y(3, 2) \approx \frac{f(3, 2.01) - f(3, 2)}{0.01} = \frac{\frac{3^2}{(2.01+1)} - \frac{3^2}{(2+1)}}{0.01} = -0.99668.$$

With $h = 0.0001$, we get

$$f_y(3, 2) \approx \frac{f(3, 2.0001) - f(3, 2)}{0.0001} = \frac{\frac{3^2}{(2.0001+1)} - \frac{3^2}{(2+1)}}{0.0001} = -0.9999667.$$

Thus, it seems that the difference quotient is approaching -1, so we estimate

$$f_y(3, 2) \approx -1.$$

13. We can use the formula $\Delta f \approx \Delta x f_x + \Delta y f_y$. Applying this formula in order to estimate $f(105, 21)$ from the known value of $f(100, 20)$ gives

$$f(105, 21) \approx f(100, 20) + (105 - 100)f_x(100, 20) + (21 - 20)f_y(100, 20)$$
$$= 2750 + 5 \cdot 4 + 1 \cdot 7$$
$$= 2777.$$

17. Since the average rate of change of the temperature adjusted for wind-chill is about -2.6 (drops by $2.6°$F), with every 1 mph increase in wind speed from 5 mph to 10 mph, when the true temperature stays constant at $20°$F, we know that

$$f_w(5, 20) \approx -2.6.$$

21. (a) The partial derivative $f_x = 350$ tells us that R increases by \$350 as x increases by 1. Thus, $f(201, 400) = f(200, 400) + 350 = 150,000 + 350 = 150,350$.
    (b) The partial derivative $f_y = 200$ tells us that R increases by \$200 as y increases by 1. Since y is increasing by 5, we have $f(200, 405) = f(200, 400) + 5(200) = 150,000 + 1,000 = 151,000$.
    (c) Here x is increasing by 3 and y is increasing by 6. We have $f(203, 406) = f(200, 400) + 3(350) + 6(200) = 150,000 + 1,050 + 1,200 = 152,250$.

    In this problem, the partial derivatives gave exact results, but in general they only give an estimate of the changes in the function.

25. Estimating from the contour diagram, using positive increments for Δx and Δy, we have, for point A,

$$\left.\frac{\partial n}{\partial x}\right|_{(A)} \approx \frac{1.5 - 1}{67 - 59} = \frac{1/2}{8} = \frac{1}{16} \approx 0.06 \; \frac{\text{foxes/km}^2}{\text{km}}$$

$$\left.\frac{\partial n}{\partial y}\right|_{(A)} \approx \frac{0.5 - 1}{60 - 51} = -\frac{1/2}{9} = -\frac{1}{18} \approx -0.06 \; \frac{\text{foxes/km}^2}{\text{km}}.$$

So, from point A the fox population density increases as we move eastward. The population density decreases as we move north from A.

At point B,

$$\left. \frac{\partial n}{\partial x} \right|_{(B)} \approx \frac{0.75 - 1}{135 - 115} = -\frac{1/4}{20} = -\frac{1}{80} \approx -0.01 \quad \frac{\text{foxes/km}^2}{\text{km}}$$

$$\left. \frac{\partial n}{\partial y} \right|_{(B)} \approx \frac{0.5 - 1}{120 - 110} = -\frac{1/2}{10} = -\frac{1}{20} \approx -0.05 \quad \frac{\text{foxes/km}^2}{\text{km}}.$$

So, fox population density decreases as we move both east and north of B. However, notice that the partial derivative $\partial n / \partial x$ at B is smaller in magnitude than the others. Indeed if we had taken a negative Δx we would have obtained an estimate of the opposite sign. This suggests that better estimates for B are

$$\left. \frac{\partial n}{\partial x} \right|_{(B)} \approx 0 \quad \frac{\text{foxes/km}^2}{\text{km}}$$

$$\left. \frac{\partial n}{\partial y} \right|_{(B)} \approx -0.05 \quad \frac{\text{foxes/km}^2}{\text{km}}.$$

At point C,

$$\left. \frac{\partial n}{\partial x} \right|_{(C)} \approx \frac{2 - 1.5}{135 - 115} = \frac{1/2}{20} = \frac{1}{40} \approx 0.02 \quad \frac{\text{foxes/km}^2}{\text{km}}$$

$$\left. \frac{\partial n}{\partial y} \right|_{(C)} \approx \frac{2 - 1.5}{80 - 55} = \frac{1/2}{25} = \frac{1}{50} \approx 0.02 \quad \frac{\text{foxes/km}^2}{\text{km}}.$$

So, the fox population density increases as we move east and north of C. Again, if these estimates were made using negative values for Δx and Δy we would have had estimates of the opposite sign. Thus, better estimates are

$$\left. \frac{\partial n}{\partial x} \right|_{(C)} \approx 0 \quad \frac{\text{foxes/km}^2}{\text{km}}$$

$$\left. \frac{\partial n}{\partial y} \right|_{(C)} \approx 0 \quad \frac{\text{foxes/km}^2}{\text{km}}.$$

Solutions for Section 7.4

1.

$$f(1, 2) = (1)^3 + 3(2)^2 = 13$$
$$f_x(x, y) = 3x^2 + 0 \Rightarrow f_x(1, 2) = 3(1)^2 = 3$$
$$f_y(x, y) = 0 + 6y \Rightarrow f_y(1, 2) = 6(2) = 12$$

5. $f_x(x, y) = 4x + 0 = 4x$
    $f_y(x, y) = 0 + 6y = 6y$

9. $f_x(x, y) = 2 \cdot 100xy = 200xy$
    $f_y(x, y) = 100x^2 \cdot 1 = 100x^2$

13. $\dfrac{\partial A}{\partial h} = \dfrac{1}{2}(a + b)$

17.

$$f(500, 1000) = 16 + 1.2(500) + 1.5(1000) + 0.2(500)(1000)$$
$$= \$102,116$$

The cost of producing 500 units of item 1 and 1000 units of item 2 is \$102,116.

$$f_{q_1}(q_1, q_2) = 0 + 1.2 + 0 + 0.2q_2$$

So $f_{q_1}(500, 1000) = 1.2 + 0.2(1000) = \$201.20$ per unit. When the company is producing at $q_1 = 500$, $q_2 = 1000$, the cost of producing one more unit of item 1 is \$201.20.

$$f_{q_2}(q_1, q_2) = 0 + 0 + 1.5 + 0.2q_1$$

So $f_{q_2}(500, 1000) = 1.5 + 0.2(500) = \$101.50$ per unit. When the company is producing at $q_1 = 500$, $q_2 = 1000$, the cost of producing one more unit of item 2 is \$101.50.

21. (a) $Q_K = 18.75K^{-0.25}L^{0.25}$, $Q_L = 6.25K^{0.75}L^{-0.75}$.
    (b) When $K = 60$ and $L = 100$,

$$Q = 25 \cdot 60^{0.75} \cdot 100^{0.25} = 1704.33$$
$$Q_K = 18.75 \cdot 60^{-0.25} 100^{0.25} = 21.3$$
$$Q_L = 6.25 \cdot 60^{0.75} 100^{-0.75} = 4.26$$

   (c) Q is actual quantity being produced. Q_K is how much more could be produced if you increased K by one unit. Q_L is how much more could be produced if you increased L by 1.

25. $f_x = 2xy$ and $f_y = x^2$, so $f_{xx} = 2y$, $f_{xy} = 2x$, $f_{yy} = 0$ and $f_{yx} = 2x$.

29. $f_x = 2xy^2$ and $f_y = 2x^2y$, so $f_{xx} = 2y^2$, $f_{xy} = 4xy$, $f_{yy} = 2x^2$ and $f_{yx} = 4xy$.

33. $P_K = 2L^2$ and $P_L = 4KL$, so $P_{KK} = 0$, $P_{LL} = 4K$ and $P_{KL} = P_{LK} = 4L$.

37. Since $f_x(x, y) = 4x^3y^2 - 3y^4$, we could have

$$f(x, y) = x^4y^2 - 3xy^4.$$

In that case,

$$f_y(x, y) = \frac{\partial}{\partial y}(x^4y^2 - 3xy^4) = 2x^4y - 12xy^3$$

as expected. More generally, we could have $f(x, y) = x^4y^2 - 3xy^4 + C$, where C is any constant.

Solutions for Section 7.5

1. Mississippi lies entirely within a region designated as 80s so we expect both the maximum and minimum daily high temperatures within the state to be in the 80s. The southwestern-most corner of the state is close to a region designated as 90s, so we would expect the temperature here to be in the high 80s, say 87-88. The northern-most portion of the state is located near the center of the 80s region. We might expect the high temperature there to be between 83-87.

   Alabama also lies completely within a region designated as 80s so both the high and low daily high temperatures within the state are in the 80s. The southeastern tip of the state is close to a 90s region so we would expect the temperature here to be about 88-89 degrees. The northern-most part of the state is near the center of the 80s region so the temperature there is 83-87 degrees.

   Pennsylvania is also in the 80s region, but it is touched by the boundary line between the 80s and a 70s region. Thus we expect the low daily high temperature to occur there and be about 80 degrees. The state is also touched by a boundary line of a 90s region so the high will occur there and be 89-90 degrees.

   New York is split by a boundary between an 80s and a 70s region, so the northern portion of the state is likely to be about 74-76 while the southern portion is likely to be in the low 80s, maybe 81-84 or so.

   California contains many different zones. The northern coastal areas will probably have the daily high as low as 65-68, although without another contour on that side, it is difficult to judge how quickly the temperature is dropping off to the west. The tip of Southern California is in a 100s region, so there we expect the daily high to be 100-101.

   Arizona will have a low daily high around 85-87 in the northwest corner and a high in the 100s, perhaps 102-107 in its southern regions.

   Massachusetts will probably have a high daily high around 81-84 and a low daily high of 70.

5. At a critical point $f_x = 2x + y = 0$ and $f_y = x + 3 = 0$, so $(-3, 6)$ is the only critical point. Since $f_{xx}f_{yy} - f_{xy}^2 = -1 < 0$, the point $(-3, 6)$ is neither a local maximum nor a local minimum.

9. At a local maximum value of f,
$$\frac{\partial f}{\partial x} = -2x - B = 0.$$
We are told that this is satisfied by $x = -2$. So $-2(-2) - B = 0$ and $B = 4$. In addition,
$$\frac{\partial f}{\partial y} = -2y - C = 0$$
and we know this holds for $y = 1$, so $-2(1) - C = 0$, giving $C = -2$. We are also told that the value of f is 15 at the point $(-2, 1)$, so
$$15 = f(-2, 1) = A - ((-2)^2 + 4(-2) + 1^2 - 2(1)) = A - (-5), \text{ so } A = 10.$$

Now we check that these values of A, B, and C give $f(x, y)$ a local maximum at the point $(-2, 1)$. Since
$$f_{xx}(-2, 1) = -2,$$
$$f_{yy}(-2, 1) = -2$$
and
$$f_{xy}(-2, 1) = 0,$$
we have that $f_{xx}(-2, 1)f_{yy}(-2, 1) - f_{xy}^2(-2, 1) = (-2)(-2) - 0 > 0$ and $f_{xx}(-2, 1) < 0$. Thus, f has a local maximum value 15 at $(-2, 1)$.

Solutions for Section 7.6

1. We wish to optimize $f(x, y) = xy$ subject to the constraint $g(x, y) = 5x + 2y = 100$. To do this we must solve the following system of equations:
$$f_x(x, y) = \lambda g_x(x, y), \qquad \text{so } y = 5\lambda$$
$$f_y(x, y) = \lambda g_y(x, y), \qquad \text{so } x = 2\lambda$$
$$g(x, y) = 100, \qquad \text{so } 5x + 2y = 100$$
We substitute in the third equation to obtain $5(2\lambda) + 2(5\lambda) = 100$, so $\lambda = 5$. Thus,
$$x = 10 \quad y = 25 \quad \lambda = 5$$
corresponding to optimal $f(x, y) = (10)(25) = 250$.

5. Our objective function is $f(x, y) = x + y$ and our equation of constraint is $g(x, y) = x^2 + y^2 = 1$. To optimize $f(x, y)$ with Lagrange multipliers, we solve the following system of equations
$$f_x(x, y) = \lambda g_x(x, y), \qquad \text{so } 1 = 2\lambda x$$
$$f_y(x, y) = \lambda g_y(x, y), \qquad \text{so } 1 = 2\lambda y$$
$$g(x, y) = 1, \qquad \text{so } x^2 + y^2 = 1$$
Solving for λ gives
$$\lambda = \frac{1}{2x} = \frac{1}{2y},$$
which tells us that $x = y$. Going back to our equation of constraint, we use the substitution $x = y$ to solve for y:
$$g(y, y) = y^2 + y^2 = 1$$
$$2y^2 = 1$$
$$y^2 = \frac{1}{2}$$
$$y = \pm\sqrt{\frac{1}{2}} = \pm\frac{\sqrt{2}}{2}.$$
Since $x = y$, our critical points are $(\frac{\sqrt{2}}{2}, \frac{\sqrt{2}}{2})$ and $(-\frac{\sqrt{2}}{2}, -\frac{\sqrt{2}}{2})$. Evaluating f at these points we find that the maximum value is $f(\frac{\sqrt{2}}{2}, \frac{\sqrt{2}}{2}) = \sqrt{2}$ and the minimum value is $f(-\frac{\sqrt{2}}{2}, -\frac{\sqrt{2}}{2}) = -\sqrt{2}$.

9. (a) To be producing the maximum quantity Q under the cost constraint given, the firm should be using K and L values given by

$$\frac{\partial Q}{\partial K} = 0.6aK^{-0.4}L^{0.4} = 20\lambda$$

$$\frac{\partial Q}{\partial L} = 0.4aK^{0.6}L^{-0.6} = 10\lambda$$

$$20K + 10L = 150.$$

Hence $\dfrac{0.6aK^{-0.4}L^{0.4}}{0.4aK^{0.6}L^{-0.6}} = 1.5\dfrac{L}{K} = \dfrac{20\lambda}{10\lambda} = 2$, so $L = \dfrac{4}{3}K$. Substituting in $20K + 10L = 150$, we obtain $20K + 10\left(\dfrac{4}{3}\right)K = 150$. Then $K = \dfrac{9}{2}$ and $L = 6$, so capital should be reduced by $\dfrac{1}{2}$ unit, and labor should be increased by 1 unit.

(b) $\dfrac{\text{New production}}{\text{Old production}} = \dfrac{a4.5^{0.6}6^{0.4}}{a5^{0.6}5^{0.4}} \approx 1.01$, so tell the board of directors, "Reducing the quantity of capital by 1/2 unit and increasing the quantity of labor by 1 unit will increase production by 1% while holding costs to \$150."

13. (a) We wish to maximize q subject to the constraint that cost $C = 10W + 20K = 3000$. To optimize q according to this, we must solve the following system of equations:

$$q_W = \lambda C_W, \qquad \text{so } \frac{9}{2}W^{-\frac{1}{4}}K^{\frac{1}{4}} = \lambda 10$$

$$q_K = \lambda C_K, \qquad \text{so } \frac{3}{2}W^{\frac{3}{4}}K^{-\frac{3}{4}} = \lambda 20$$

$$C = 3000, \qquad \text{so } 10W + 20K = 3000$$

Dividing yields $K = \frac{1}{6}W$, so substituting into C gives

$$10W + 20\left(\frac{1}{6}W\right) = \frac{40}{3}W = 3000.$$

Thus $W = 225$ and $K = 37.5$. Substituting both answers to find λ gives

$$\lambda = \frac{\frac{9}{2}(225)^{-\frac{1}{4}}(37.5)^{\frac{1}{4}}}{10} = 0.2875.$$

We also find the optimum quantity produced, $q = 6(225)^{\frac{3}{4}}(37.5)^{\frac{1}{4}} = 862.57$.

(b) When the budget is increased by one dollar, we substitute the relation $K_1 = \frac{1}{6}W_1$ into $10W_1 + 20K_1 = 3001$ which gives $10W_1 + 20(\frac{1}{6}W_1) = \frac{40}{3}W_1 = 3001$. Solving yields $W_1 = 225.075$ and $K_1 = 37.513$, so $q_1 = 862.86 = q + 0.29$. Thus production has increased by $0.29 \approx \lambda$, the Lagrange Multiplier.

17. (a) The curves are shown in Figure 7.10.

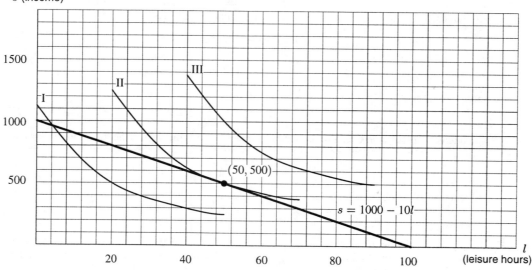

Figure 7.10

(b) The income equals $10/hour times the number of hours of work:

$$s = 10(100 - l) = 1000 - 10l.$$

(c) The graph of this constraint is the straight line in Figure 7.10.

(d) For any given salary, curve III allows for the most leisure time, curve I the least. Similarly, for any amount of leisure time, curve III also has the greatest salary, and curve I the least. Thus, any point on curve III is preferable to any point on curve II, which is preferable to any point on curve I. We prefer to be on the outermost curve that our constraint allows. We want to choose the point on $s = 1000 - 10l$ which is on the most preferable curve. Since all the curves are concave up, this occurs at the point where $s = 1000 - 10l$ is *tangent* to curve II. So we choose $l = 50$, $s = 500$, and work 50 hours a week.

Solutions for Chapter 7 Review

1. (a)

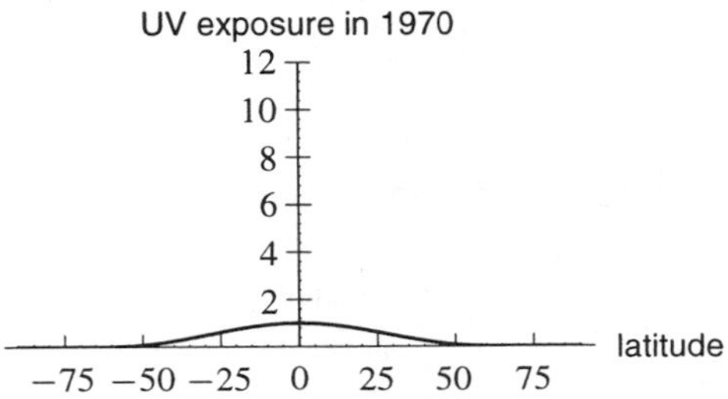

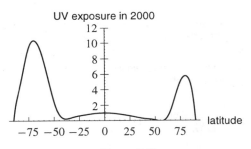

Figure 7.11 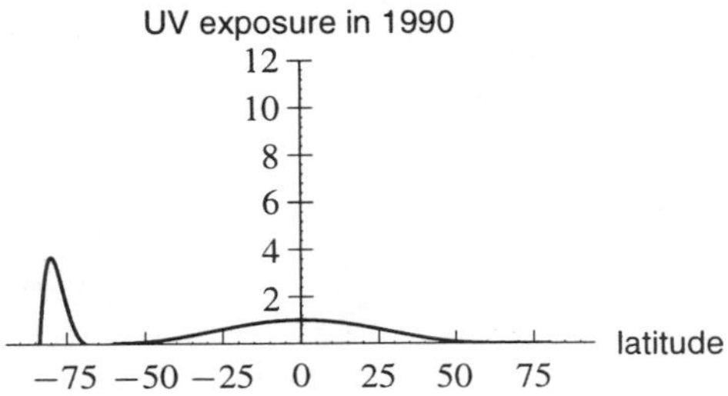   **Figure 7.12**

Figure 7.13

(b)

TABLE 7.5 *Latitude of most severe exposure*

Year	1970	1975	1980	1985	1990	1995	2000	2005	2010
Latitude	0	0	0	0	-80	-70	-70	-70	-60

(c) The latitude of most severe exposure was $-80°$ in 1990 (near the south pole) and moves to $-70°$ and then $-60°$. The reason of this is that a large hole in the ozone is forming above the south pole, and a smaller hole is opening above the north pole. These holes in the ozone allow more UV light to pass through the atmosphere to the surface of the planet.

5. At any fixed position, as time passes, the fallout increases, although increasing more slowly as time passes. Also, the fallout decreases with distance. This answer is as we would expect: the fallout increases with time but decreases with distance from the volcano.

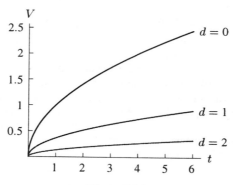

Figure 7.14

9. (a)

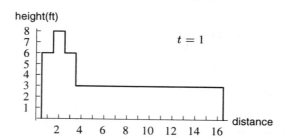

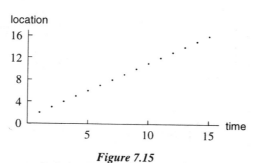

Figure 7.15

(c) The "wave" at a sports arena.

13. $\dfrac{\partial Q}{\partial p_1} = 50p_2$, $\dfrac{\partial Q}{\partial p_2} = 50p_1 - 2p_2$.

17. To read off the cross-sections of f with t fixed, we choose a t value and move horizontally across the diagram looking at the values on the contours. For $t = 0$, as we move from the left at $x = 0$ to the right at $x = \pi$, we cross contours of 0.25, 0.50, 0.75 and reach a maximum at $x = \pi/2$, and then decrease back to 0. That is because if time is fixed at $t = 0$, then $f(x, 0)$ is the displacement of the string at that time: no displacement at $x = 0$ and $x = \pi$ and greatest displacement at $x = \pi/2$. For cross-sections with t fixed at larger values, as we move along a horizontal line, we cross fewer contours and reach a smaller maximum value: the string is becoming less curved. At time $t = \pi/2$, the string is straight so we see a value of 0 all the way across the diagram, namely a contour with value 0. For $t = \pi$, the string has vibrated to the other side and the displacements are negative as we read across the diagram reaching a minimum at $x = \pi/2$.

The cross-sections of f with x fixed are read vertically. At $x = 0$ and $x = \pi$, we see vertical contours of value 0 because the end points of the string have 0 displacement no matter what time it is. The cross-section for $x = \pi/2$ is found by moving vertically up the diagram at $x = \pi/2$. As we expect, the contour values are largest at $t = 0$, zero at $t = \pi/2$, and a minimum at $t = \pi$.

Notice that the spacing of the contours is also important. For example, for the $t = 0$ cross-section, contours are most closely spaced at the end points at $x = 0$ and $x = \pi$ and most spread out at $x = \pi/2$. That is because the shape of the string at time $t = 0$ is a sine curve, which is steepest at the end points and relatively flat in the middle. Thus, the contour diagram shows the steepest terrain at the end points and flattest terrain in the middle.

21. The sign of $\partial f/\partial P_1$ tells you whether f (the number of people who ride the bus) increases or decreases when P_1 is increased. Since P_1 is the price of taking the bus, as it increases, f should decrease. This is because fewer people will be willing to pay the higher price, and more people will choose to ride the train. On the other hand, the sign of $\dfrac{\partial f}{\partial P_2}$ tells you the change in f as P_2 increases. Since P_2 is the cost of riding the train, as it increases, f should increase. This is because fewer people will be willing to pay the higher fares for the train, and more people will choose to ride the bus.

Therefore, $\dfrac{\partial f}{\partial P_1} < 0$ and $\dfrac{\partial f}{\partial P_2} > 0$.

25. Since $\dfrac{\partial f}{\partial y}$ and $\dfrac{\partial f}{\partial y}$ are defined everywhere, a critical point will occur where $\dfrac{\partial f}{\partial x} = 0$ and $\dfrac{\partial f}{\partial y} = 0$. So:

$$\frac{\partial f}{\partial x} = 2x - 4 = 0 \Rightarrow x = 2$$

$$\frac{\partial f}{\partial y} = 6y + 6 = 0 \Rightarrow y = -1$$

$(2, -1)$ is the critical point of $f(x, y)$.

29. (a) About 15 feet along the wall, because that's where there are regions of cold air ($55°$F and $65°$F).

    (b) Roughly between 10 am and 12 noon, and between 4 pm and 6 pm.

    (c) Roughly between midnight and 2 am, between 10 am and 1 pm, and between 4 pm and 9 pm, since that is when the temperature near the heater is greater than $80°$F.

    (d)

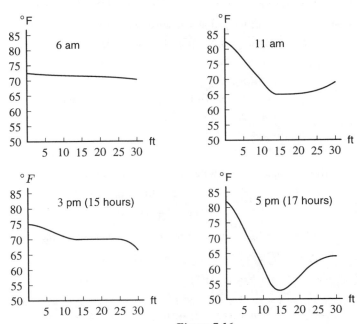

Figure 7.16

(e)

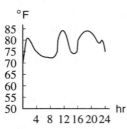

Figure 7.17: Temp. vs.
Time at heater

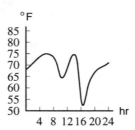

Figure 7.18: Temp. vs.
Time at window

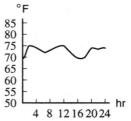

Figure 7.19: Temp. vs.
Time midway between
heater and window

(f)   The temperature at the window is colder at 5 pm than at 11 am because the outside temperature is colder at 5 pm than at 11 am.

(g)   The thermostat is set to roughly 70°F. We know this because the temperature in the room stays close to 70°F until we get close (a couple of feet) to the window.

(h)   We are told that the thermostat is about 2 feet from the window. Thus, the thermostat is either about 13 feet or about 17 feet from the wall. If the thermostat is set to 70°F, every time the temperature at the thermostat goes over or under 70°F, the heater turns off or on. Look at the point at which the vertical lines at 13 feet or about 17 feet cross the 70°F contours. We need to decide which of these crossings correspond best with the times that the heater turns on and off. (These times can be seen along the wall.) Notice that the 17 foot line does not cross the 70°F contour after 16 hours (4 pm). Thus, if the thermostat were 17 feet from the wall, the heater would not turn off after 4 pm. However, the heater does turn off at about 21 hours (9 pm). Since this is the time that the 13 foot line crosses the 70°F contour, we estimate that the thermostat is about 13 feet away from the wall.

Solutions to Problems on Deriving the Formula for Regression Lines

1.   Let the line be in the form $y = b + mx$. When x equals -1, 0 and 1, then y equals $b - m$, b, and $b + m$, respectively. The sum of the squares of the vertical distances, which is what we want to minimize, is

$$f(m, b) = (2 - (b - m))^2 + (-1 - b)^2 + (1 - (b + m))^2.$$

To find the critical points, we compute the partial derivatives with respect to m and b,

$$\begin{aligned} f_m &= 2(2 - b + m) + 0 + 2(1 - b - m)(-1) \\ &= 4 - 2b + 2m - 2 + 2b + 2m \\ &= 2 + 4m, \\ f_b &= 2(2 - b + m)(-1) + 2(-1 - b)(-1) + 2(1 - b - m)(-1) \\ &= -4 + 2b - 2m + 2 + 2b - 2 + 2b + 2m \\ &= -4 + 6b. \end{aligned}$$

Setting both partial derivatives equal to zero, we get a system of equations:

$$2 + 4m = 0,$$
$$-4 + 6b = 0.$$

The solution is $m = -1/2$ and $b = 2/3$. You can check that it is a minimum. Hence, the regression line is $y = \dfrac{2}{3} - \dfrac{1}{2}x$.

5.   We have $\sum x_i = 6$, $\sum y_i = 5$, $\sum x_i^2 = 14$, and $\sum y_i x_i = 12$. Thus

$$\begin{aligned} b &= \big((14)(5) - (6)(12)\big) \big/ \big((3)(14) - (6^2)\big) \\ &= -2/6 = -1/3 \\ m &= \big((3)(12) - (6)(5)\big) \big/ \big((3)(14) - (6^2)\big) \\ &= 6/6 = 1. \end{aligned}$$

The line is $y = x - \dfrac{1}{3}$, which agrees with the answer to Example 1.

CHAPTER EIGHT

Solutions for Section 8.1

1. Since $y = t^4$, the derivative is $dy/dt = 4t^3$. We have

$$\text{Left-side} = t\frac{dy}{dt} = t(4t^3) = 4t^4.$$

$$\text{Right-side} = 4y = 4t^4.$$

Since the substitution $y = t^4$ makes the differential equation true, $y = t^4$ is in fact a solution.

5. We know that at time $t = 0$, the value of y is 8. Since we are told that $dy/dt = 0.5y$, we know that at time $t = 0$

$$\frac{dy}{dt} = 0.5(8) = 4.$$

As t goes from 0 to 1, y will increase by 4, so at $t = 1$,

$$y = 8 + 4 = 12.$$

Likewise, we get that at $t = 1$,

$$\frac{dy}{dt} = .5(12) = 6$$

so that at $t = 2$,

$$y = 12 + 6 = 18.$$

At $t = 2$, $\dfrac{dy}{dt} = .5(18) = 9$ so that at $t = 3$, $y = 18 + 9 = 27$.

At $t = 3$, $\dfrac{dy}{dt} = .5(27) = 13.5$ so that at $t = 4$, $y = 27 + 13.5 = 40.5$.

Thus we get the following table

t	0	1	2	3	4
y	8	12	18	27	40.5

9. (a) = (III), (b) = (IV), (c) = (I), (d) = (II).

13. Since $y = x^2 + k$ we know that
$$y' = 2x.$$

Substituting $y = x^2 + k$ and $y' = 2x$ into the differential equation we get

$$\begin{aligned}
10 = 2y - xy' \\
= 2(x^2 + k) - x(2x) \\
= 2x^2 + 2k - 2x^2 \\
= 2k
\end{aligned}$$

Thus, $k = 5$ is the only solution.

17. We first compute dy/dx for each of the functions on the right.

If $y = x^3$ then

$$\frac{dy}{dx} = 3x^2$$
$$= 3\frac{y}{x}.$$

If $y = 3x$ then

$$\frac{dy}{dx} = 3$$
$$= \frac{y}{x}.$$

If $y = e^{3x}$ then

$$\frac{dy}{dx} = 3e^{3x}$$
$$= 3y.$$

If $y = 3e^x$ then

$$\frac{dy}{dx} = 3e^x$$
$$= y.$$

Finally, if $y = x$ then

$$\frac{dy}{dx} = 1$$
$$= \frac{y}{x}.$$

Comparing our calculated derivatives with the right-hand sides of the differential equations we see that (a) is solved by (II) and (V), (b) is solved by (I), (c) is not solved by any of our functions, (d) is solved by (IV) and (e) is solved by (III).

Solutions for Section 8.2

1. (a)

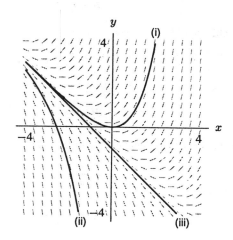

(b) The solution through $(-1, 0)$ appears to be linear with equation $y = -x - 1$.

(c) If $y = -x - 1$, then $y' = -1$ and $x + y = x + (-x - 1) = -1$, so this checks as a solution.

5.

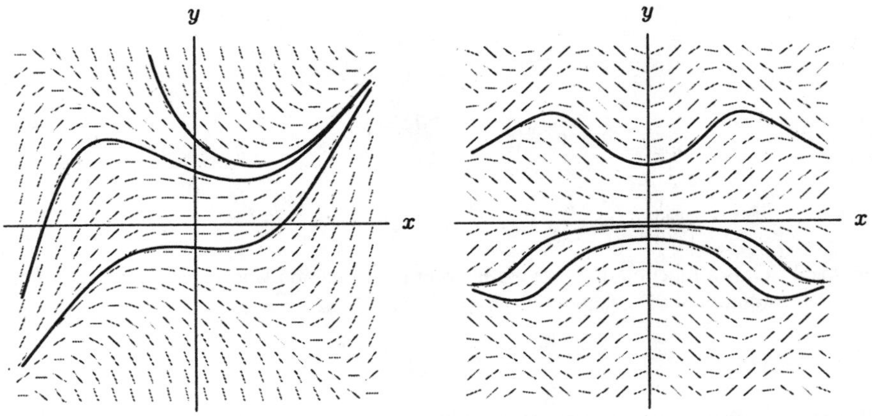

Figure 8.1

Other choices of solution curves are, of course, possible.

9. As $x \to \infty$, $y \to \infty$, no matter what the starting point is.

13. When $a = 1$ and $b = 2$, the Gompertz equation is $y' = -y \ln(y/2) = y \ln(2/y) = y(\ln 2 - \ln y)$. This differential equation is similar to the differential equation $y' = y(2 - y)$ in certain ways. For example, in both equations y' is positive for $0 < y < 2$ and negative for $y > 2$. Also, for y-values close to 2, $(\ln 2 - \ln y)$ and $(2 - y)$ are both close to 0, so $y(\ln 2 - \ln y)$ and $y(2 - y)$ are approximately equal to zero. Thus around $y = 2$ the slope fields look almost the same. This happens again around $y = 0$, since around $y = 0$ both $y(2 - y)$ and $y(\ln 2 - \ln y)$ go to 0. Finally, for $y > 2$, $\ln y$ grows much slower than y, so the slope field for $y' = y(\ln 2 - \ln y)$ is less steep, negatively, than for $y' = y(2 - y)$.

Solutions for Section 8.3

1. The equation given is in the form

$$\frac{dP}{dt} = kP.$$

Thus we know that the general solution to this equation will be

$$P = Ce^{kt}.$$

And in our case, with $k = 0.02$ and $C = 20$ we get

$$P = 20e^{0.02t}.$$

5. Rewriting we get

$$\frac{dy}{dx} = -\frac{1}{3}y.$$

We know that the general solution to an equation in the form

$$\frac{dy}{dx} = ky$$

is

$$y = Ce^{kx}.$$

Thus in our case we get

$$y = Ce^{-\frac{1}{3}x}.$$

We are told that $y(0) = 10$ so we get

$$y(x) = Ce^{-\frac{1}{3}x}$$
$$y(0) = 10 = Ce^0$$
$$C = 10$$

Thus we get

$$y = 10e^{-\frac{1}{3}x}.$$

9. Since the right hand side is a function of x, we are looking for a function whose derivative is $5x$. One such function is $5x^2/2$, so the general solution of the equation $dB/dx = 5x$ is

$$B = \frac{5}{2}x^2 + C$$

which is neither an exponentially growing nor an exponentially decaying function.

13. (a) We know that the balance, B, increases at a rate proportional to the current balance. Since interest is being earned at a rate of 7% compounded continuously we have

$$\text{Rate at which interest is earned} = 7\% \text{ (Current balance)}$$

or in other words, if t is time in years,

$$\frac{dB}{dt} = 7\%(B) = 0.07B.$$

(b) The equation is in the form

$$\frac{dB}{dt} = kB$$

so we know that the general solution will be

$$B = B_0 e^{kt}$$

where B_0 is the value of B when $t = 0$, i.e., the initial balance. In our case we have $k = 0.07$ so we get

$$B = B_0 e^{0.07t}.$$

(c) We are told that the initial balance, B_0, is \$5000 so we get

$$B = 5000e^{0.07t}.$$

(d) Substituting the value $t = 10$ into our formula for B we get

$$B = 5000e^{0.07t}$$
$$B(10) = 5000e^{0.07(10)}$$
$$= 5000e^{0.7}$$
$$B(10) \approx \$10,068.75$$

17. Lake Superior will take the longest, because the lake is largest (V is largest) and water is moving through it most slowly (r is smallest). Lake Erie looks as though it will take the least time because V is smallest and r is close to the largest. For Erie, $k = r/V = 175/460 \approx 0.38$. The lake with the largest value of r is Ontario, where $k = r/V = 209/1600 \approx 0.13$. Since e^{-kt} decreases faster for larger k, Lake Erie will take the shortest time for any fixed fraction of the pollution to be removed.

For Lake Superior,

$$\frac{dQ}{dt} = -\frac{r}{V}Q = -\frac{65.2}{12,200}Q \approx -0.0053Q$$

so

$$Q = Q_0 e^{-0.0053t}.$$

When 80% of the pollution has been removed, 20% remains so $Q = 0.2Q_0$. Substituting gives us

$$0.2Q_0 = Q_0 e^{-0.0053t}$$

so

$$t = -\frac{\ln(0.2)}{0.0053} \approx 301 \text{ years.}$$

(Note: The 301 is obtained by using the exact value of $\frac{r}{V} = \frac{65.2}{12,200}$, rather than 0.0053. Using 0.0053 gives 304 years.)
For Lake Erie, as in the text

$$\frac{dQ}{dt} = -\frac{r}{V}Q = -\frac{175}{460}Q \approx -0.38Q$$

so

$$Q = Q_0 e^{-0.38t}.$$

When 80% of the pollution has been removed

$$0.2Q_0 = Q_0 e^{-0.38t}$$

$$t = -\frac{\ln(0.2)}{0.38} \approx 4 \text{ years.}$$

So the ratio is

$$\frac{\text{Time for Lake Superior}}{\text{Time for Lake Erie}} \approx \frac{301}{4} \approx 75.$$

In other words it will take about 75 times as long to clean Lake Superior as Lake Erie.

21. (a) Since the rate of change is proportional to the amount present, $dy/dt = ky$ for some constant k.
    (b) Solving the differential equation, we have $y = Ae^{kt}$, where A is the initial amount. Since 100 grams become 54.9 grams in one hour, $54.9 = 100e^k$, so $k = \ln(54.9/100) \approx -0.5997$.
    Thus, after 10 hours, there remains $100e^{(-0.5997)10} \approx 0.2486$ grams.

Solutions for Section 8.4

1. We know that the general solution to a differential equation of the form

$$\frac{dH}{dt} = k(H - A)$$

is

$$H = A + Ce^{kt}.$$

Thus in our case we get

$$H = 75 + Ce^{3t}.$$

We know that at $t = 0$ we have $H = 0$, so solving for C we get

$$H = 75 + Ce^{3t}$$
$$0 = 75 + Ce^{3(0)}$$
$$-75 = Ce^0$$
$$C = -75.$$

Thus we get

$$H = 75 - 75e^{3t}.$$

5. We know that the general solution to a differential equation of the form

$$\frac{dQ}{dt} = k(Q - A)$$

is

$$H = A + Ce^{kt}.$$

To get our equation in this form we factor out a 0.3 to get

$$\frac{dQ}{dt} = 0.3\left(Q - \frac{120}{0.3}\right) = 0.3(Q - 400).$$

Thus in our case we get

$$Q = 400 + Ce^{0.3t}.$$

We know that at $t = 0$ we have $Q = 50$, so solving for C we get

$$Q = 400 + Ce^{0.3t}$$
$$50 = 400 + Ce^{0.3(0)}$$
$$-350 = Ce^0$$
$$C = -350.$$

Thus we get

$$Q = 400 - 350e^{0.3t}.$$

9. In order to check that $y = A + Ce^{kt}$ is a solution to the differential equation

$$\frac{dy}{dt} = k(y - A),$$

we must show that the derivative of y with respect to t is equal to $k(y - A)$:

$$y = A + Ce^{kt}$$
$$\frac{dy}{dt} = 0 + (Ce^{kt})(k)$$
$$= kCe^{kt}$$
$$= k(Ce^{kt} + A - A)$$
$$= k\left((Ce^{kt} + A) - A\right)$$
$$= k(y - A)$$

13. (a) We know that the equilibrium solution are the functions satisfying the differential equation whose derivative everywhere is 0. Thus we must solve the equation

$$\frac{dy}{dt} = 0.$$

Solving we get

$$\frac{dy}{dt} = 0$$
$$0.2(y - 3)(y + 2) = 0$$
$$(y - 3)(y + 2) = 0$$

Thus the solutions are $y = 3$ and $y = -2$.

(b) Looking at Figure 8.2 we see that the line $y = 3$ is an unstable solution while the line $y = -2$ is a stable solution.

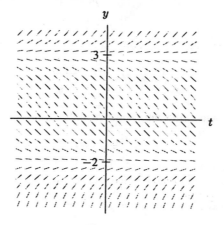

Figure 8.2

17. (a) $\dfrac{dT}{dt} = -k(T - A)$, where $A = 68°$F is the temperature of the room.

(b) We know that the general solution to a differential equation of the form

$$\frac{dT}{dt} = -k(T - 68)$$

is

$$T = Ce^{-kt} + 68.$$

We know that the temperature of the body is $90.3°$F at 9 am. Thus, letting $t = 0$ correspond to 9 am, we get

$$T = Ce^{-kt} + 68$$
$$T(0) = 90.3 = Ce^{-k(0)} + 68$$
$$90.3 = Ce^0 + 68$$
$$C = 90.3 - 68 = 22.3$$

Thus

$$T = 68 + 22.3e^{-kt}.$$

At $t = 1$, we have

$$89.0 = 68 + 22.3e^{-k}$$
$$21 = 22.3e^{-k}$$
$$k = -\ln\frac{21}{22.3} \approx 0.06.$$

Thus $T = 68 + 22.3e^{-0.06t}$.

We want to know when T was equal to $98.6°$F, the temperature of a live body, so

$$98.6 = 68 + 22.3e^{-0.06t}$$
$$\ln\frac{30.6}{22.3} = -0.06t$$
$$t = \left(-\frac{1}{0.06}\right)\ln\frac{30.6}{22.3}$$
$$t \approx -5.27.$$

The victim was killed approximately $5\frac{1}{4}$ hours prior to 9 am, at 3:45 am.

21. (a) $\dfrac{dy}{dt} = -k(y - a)$, where $k > 0$ and a are constants.

(b) We know that the general solution to a differential equation of the form

$$\frac{dy}{dt} = -k(y - a)$$

is

$$y = Ce^{-kt} + a.$$

We can assume that right after the course is over (at $t = 0$) 100% of the material is remembered. Thus we get

$$y = Ce^{-kt} + a$$
$$1 = Ce^0 + a$$
$$C = 1 - a.$$

Thus

$$y = (1 - a)e^{-kt} + a.$$

(c) As $t \to \infty$, $e^{-kt} \to 0$, so $y \to a$.

Thus, a represents the fraction of material which is remembered in the long run. The constant k tells us about the rate at which material is forgotten.

25. (a) The equilibrium solutions are the solutions where

$$\frac{dP}{dt} = 0$$
$$0.25P(1 - 0.0004P) = 0$$
$$P = 0, P = \frac{1}{0.0004} = 2500.$$

The equilibrium solutions are $P = 0$ and $P = 2500$.

(b)

$$\frac{dP}{dt} = 0.25P(1 - 0.0004P)$$
$$= 0.25P\left(1 - \frac{P}{\frac{1}{0.0004}}\right)$$
$$= 0.25P\left(1 - \frac{P}{2500}\right).$$

So $k = 0.25$ and $L = 2500$.

(c) The solution is of the form $P = \frac{L}{1 + Ce^{-kt}}$, so in this case

$$P = \frac{2500}{1 + Ce^{-0.25t}},$$

where C is an arbitrary constant. The long-term equilibrium population of carp in the lake is 2500.

Solutions for Section 8.5

1. Here x and y both increase at about the same rate.

5. This is an example of a predator-prey relationship. Normally, we would expect the worm population, in the absence of predators, to increase without bound. As the number of worms w increases, so would the rate of increase dw/dt; in other words, the relation $dw/dt = w$ might be a reasonable model for the worm population in the absence of predators.

   However, since there are predators (robins), dw/dt won't be that big. We must lessen dw/dt. It makes sense that the more interaction there is between robins and worms, the more slowly the worms are able to increase their numbers. Hence we lessen dw/dt by the amount wr to get $dw/dt = w - wr$. The term $-wr$ reflects the fact that more interactions between the species means slower reproduction for the worms.

   Similarly, we would expect the robin population to decrease in the absence of worms. We'd expect the population decrease at a rate related to the current population, making $dr/dt = -r$ a reasonable model for the robin population in absence of worms. The negative term reflects the fact that the greater the population of robins, the more quickly they are dying off. The wr term in $dr/dt = -r + wr$ reflects the fact that the more interactions between robins and worms, the greater the tendency for the robins to increase in population.

9. Sketching the trajectory through the point $(2, 2)$ on the slope field given shows that the maximum robin population is about 2500, and the minimum robin population is about 500. When the robin population is at its maximum, the worm population is about 1,000,000.

13. (a) Substituting $w = 2.2$ and $r = 1$ into the differential equations gives

$$\frac{dw}{dt} = 2.2 - (2.2)(1) = 0$$
$$\frac{dr}{dt} = -1 + 1(2.2) = 1.2.$$

(b) Since the rate of change of w with time is 0,

$$\text{At } t = 0.1, \text{ we estimate } w = 2.2$$

Since the rate of change of r is 1.2 thousand robins per unit time,

$$\text{At } t = 0.1, \text{ we estimate } r = 1 + 1.2(0.1) = 1.12 \approx 1.1.$$

(c)   We must recompute the derivatives. At $t = 0.1$, we have

$$\frac{dw}{dt} = 2.2 - 2.2(1.12) = -0.264$$

$$\frac{dr}{dt} = -1.12 + 1.12(2.2) = 1.344.$$

Then at $t = 0.2$, we estimate

$$w = 2.2 - 0.264(0.1) = 2.1736 \approx 2.2$$
$$r = 1.12 + 1.344(0.1) = 1.2544 \approx 1.3$$

Recomputing the derivatives at $t = 0.2$ gives

$$\frac{dw}{dt} = 2.1736 - 2.1736(1.2544) = -0.553$$

$$\frac{dr}{dt} = -1.2544 + 1.2544(2.1736) = 1.472$$

Then at $t = 0.3$, we estimate

$$w = 2.1736 - 0.553(0.1) = 2.1183 \approx 2.1$$
$$r = 1.2544 + 1.472(0.1) = 1.4016 \approx 1.4.$$

17. $\dfrac{dx}{dt} = -x + xy, \quad \dfrac{dy}{dt} = y$

21.   (a)   The x population is unaffected by the y population—it grows exponentially no matter what the y population is, even if $y = 0$. If alone, the y population decreases to zero exponentially, because its equation becomes $dy/dt = -0.1y$.

       (b)   Here, interaction between the two populations helps the y population but doesn't effect the x population. This is not a predator-prey relationship; instead, this is a one-way relationship, where the y population is helped by the existence of x's. These equations could, for instance, model the interaction of rhinocerouses (x) and dung beetles (y).

Solutions for Section 8.6

1.   Since

$$\frac{dS}{dt} = -aSI,$$

$$\frac{dI}{dt} = aSI - bI,$$

$$\frac{dR}{dt} = bI$$

we have

$$\frac{dS}{dt} + \frac{dI}{dt} + \frac{dR}{dt} = -aSI + aSI - bI + bI = 0.$$

Thus $\frac{d}{dt}(S + I + R) = 0$, so $S + I + R = $ constant.

5.   (a) $I_0 = 1, S_0 = 349$

       (b)   Since $\dfrac{dI}{dt} = 0.0026SI - 0.5I = 0.0026(349)(1) - 0.5(1) > 0$, so I is increasing. The number of infected people will increase, and the disease will spread. This is an epidemic.

9.   The threshold value of S is the value at which I is a maximum. When I is a maximum,

$$\frac{dI}{dt} = 0.04SI - 0.2I = 0,$$

so

$$S = 0.2/0.04 = 5.$$

Solutions for Chapter 8 Review

1. (a) (i) If $y = Cx^2$, then $\dfrac{dy}{dx} = C(2x) = 2Cx$. We have

$$x\frac{dy}{dx} = x(2Cx) = 2Cx^2$$

and

$$3y = 3(Cx^2) = 3Cx^2$$

Since $x\dfrac{dy}{dx} \neq 3y$, this is not a solution.

(ii) If $y = Cx^3$, then $\dfrac{dy}{dx} = C(3x^2) = 3Cx^2$. We have

$$x\frac{dy}{dx} = x(3Cx^2) = 3Cx^3,$$

and

$$3y = 3Cx^3.$$

Thus $x\dfrac{dy}{dx} = 3y$, and $y = Cx^3$ is a solution.

(iii) If $y = x^3 + C$, then $\dfrac{dy}{dx} = 3x^2$. We have

$$x\frac{dy}{dx} = x(3x^2) = 3x^3$$

and

$$3y = 3(x^3 + C) = 3x^3 + 3C.$$

Since $x\dfrac{dy}{dx} \neq 3y$, this is not a solution.

(b) The solution is $y = Cx^3$. If $y = 40$ when $x = 2$, we have

$$40 = C(2^3)$$
$$40 = C \cdot 8$$
$$C = 5.$$

5. The general solution is

$$y = Ce^{5t}.$$

9. We know that the general solution to the differential equation

$$\frac{dy}{dx} = k(y - A)$$

is

$$y = Ce^{kx} + A.$$

Thus in our case we factor out 0.2 to get

$$\frac{dy}{dx} = 0.2\left(y - \frac{8}{0.2}\right) = 0.2(y - 40).$$

Thus the general solution to our differential equation is

$$y = Ce^{0.2x} + 40,$$

where C is some constant.

13. We know that the general solution to an equation of the form

$$\frac{dP}{dt} = kP$$

is

$$P = Ce^{kt}.$$

Thus in our case the solution is

$$P = Ce^{0.08t}.$$

We know that $P = 5000$ when $t = 0$ so solving for C we get

$$P = Ce^{0.08t}$$
$$5000 = Ce^0$$
$$C = 5000.$$

Thus the solution is

$$P = 5000e^{0.08t}.$$

A graph of this function is shown in Figure 8.3.

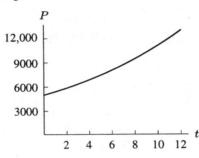

Figure 8.3

17. (a) We know that the equilibrium solutions are those functions which satisfy the differential equation and whose derivative is everywhere 0. Thus we must solve

$$0 = \frac{dy}{dx}$$
$$= 0.5y(y - 4)(2 + y)$$

Thus the equilibrium solutions are $y = 0$, $y = 4$, and $y = -2$.

(b) The slope field of the differential equation is shown in Figure 8.4. An equilibrium solution is stable if a small change in the initial conditions gives a solution which tends toward the equilibrium as the independent variable tends to positive infinity. Looking at Figure 8.4 we see that the only stable solution is $y = 0$.

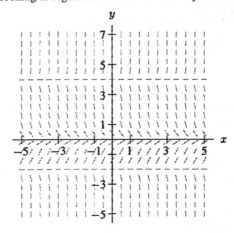

Figure 8.4

21. (a) The value of the company satisfies

Rate of change of value = Rate interest earned − Rate expenses paid

so

$$\frac{dV}{dt} = 0.02V - 80,000.$$

(b) We find V when

$$\frac{dV}{dt} = 0$$
$$0.02V - 80,000 = 0$$
$$0.02V = 80,000$$
$$V = 4,000,000$$

There is an equilibrium solution at $V = \$4,000,000$. If the company has $\$4,000,000$ in assets, its earnings will exactly equal its expenses.

(c) The general solution is

$$V = 4,000,000 + Ce^{0.02t}.$$

(d) If $V = 3,000,000$ when $t = 0$, we have $C = -1,000,000$. The solution is

$$V = 4,000,000 - 1,000,000e^{0.02t}.$$

When $t = 12$, we have

$$V = 4,000,000 - 1,000,000e^{0.02(12)}$$
$$= 4,000,000 - 1,271,249$$
$$= \$2,728,751.$$

The company is losing money.

25. Using (Rate balance changes) = (Rate interest is added)− (Rate payments are made), when the interest rate is i, we have

$$\frac{dB}{dt} = iB - 100.$$

We know that the general solution to a differential equation of the form

$$\frac{dB}{dt} = k(B - A)$$

is

$$B = Ce^{kt} + A.$$

Factoring out an i on the left side we get

$$\frac{dB}{dt} = i\left(B - \frac{100}{i}\right).$$

Thus in our case we get

$$B = Ce^{it} + \frac{100}{i}.$$

We know that at $t = 0$ we have $B = 1000$ so solving for C we get

$$1000 = Ce^0 + \frac{100}{i}$$
$$C = 1000 - \frac{100}{i}.$$

Thus $B = \frac{100}{i} + \left(1000 - \frac{100}{i}\right)e^{it}$.
When $i = 0.05$, $B = 2000 - 1000e^{0.05t}$.
When $i = 0.1$, $B = 1000$.
When $i = 0.15$, $B = 666.67 + 333.33e^{0.15t}$.
We now look at the graph when $i = 0.05$, $i = 0.1$, and $i = 0.15$.

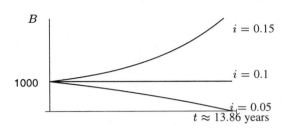

29.   (a) $dp/dt = -k(p - p_0)$, where k is constant. Notice that $k > 0$, since if $p > p_0$ then dp/dt should be negative, and if $p < p_0$ then dp/dt should be positive.

(b)   We know that the general solution to a differential equation of the form

$$\frac{dp}{dt} = -k(p - A)$$

is

$$p = Ce^{-kt} + A.$$

In our case $A = p_0$, so we get

$$p = Ce^{-kt} + p_0.$$

If I is the initial price, we get

$$p = Ce^{-kt} + p_0$$
$$I = Ce^{0} + p_0$$
$$C = I - p_0.$$

Thus we get

$$p = p_0 + (I - p_0)e^{-kt}.$$

(c)

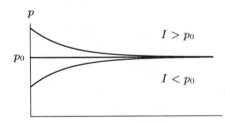

(d)   As $t \to \infty, p \to p_0$. We see this in the solution in part (b), since as $t \to \infty, e^{-kt} \to 0$. (Remember $k > 0$.) In other words, as $t \to \infty, p$ approaches the equilibrium price p_0.